SHOULDER TO SHOULDER

SHOULDER TO SHOULDER

Working Together for a Sustainable Future

EVELYN SEARLE HESS

ROWMAN & LITTLEFIELD
Lanham • Boulder • New York • London

Published by Rowman & Littlefield
An imprint of The Rowman & Littlefield Publishing Group, Inc.
4501 Forbes Boulevard, Suite 200, Lanham, Maryland 20706
www.rowman.com

6 Tinworth Street, London, SE11 5AL, United Kingdom

British Library Cataloguing in Publication Information Available

Library of Congress Cataloging-in-Publication Data

Names: Hess, Evelyn Searle, author.
Title: Shoulder to shoulder : working together for a sustainable future / Evelyn Searle Hess.
Description: Lanham : Rowman & Littlefield, [2021] | Includes bibliographical references and index. | Summary: "Shoulder to Shoulder tells the stories of five on-going environmental and social justice campaigns powered by ordinary people"—Provided by publisher.
Identifiers: LCCN 2020043347 | ISBN 9781538144398 (cloth) | ISBN 9781538144404 (ebook)
Subjects: LCSH: Environmental justice. | Social justice. | Social movements.
Classification: LCC GE220 .H47 2021 | DDC 363.7/0525—dc23
LC record available at https://lccn.loc.gov/2020043347

For Erika, Jeff, Nate, Celina, Tasha, Camila, Owen, Margaret, Evangeline, and Benjamin, and for everyone else's children, grandchildren, and great-grandchildren.

❧

In grateful memory of Sandra H. Larson, 1939–2019, who knew how to bring people together and work wholeheartedly for the common good.

❧

With gratitude for all the groups and their caring and courageous members who work daily to bring about a just and healthy world.

Contents

Acknowledgments

PEOPLE WORKING FOR THE EARTH and each other not only give me hope. They also give, and gave, me help. So many to thank. My only worry is that I will leave people out. If I do, please know I appreciate every one of you, and learned so much from you every step of the way.

I must begin with Paul Hawken, who gave me the idea, followed close behind by Hillary Johnson, who showed me where to look. Please see the bibliography for books that have been helpful plus a few other great ones on related subjects.

At all stages of the project, I was blessed with cheerleaders, readers, teachers, hand-holders, and rescuers. Those include, but are not limited to, my Kwinnim writing group: Barbara Engel, Quinton Hallett, Patty Jacobs, and Barbara Walker; Gary Jastad helped me adjust my perspective, Cecelia Hagen guided me over some early bumps; my Red Moons critique group: Grace Castle, Charleynne Gates, Cynthia Pappas, Kay Porter, and Tom Titus. I'm deeply indebted to you all. The work profited immeasurably from the reading and suggestions of Grace Castle, John Daniel, Sandra Larson, David Lawlor, Lucy Moore, and Cynthia Pappas. In addition, I'm eternally grateful to Lucy for critical connections. Huge thanks to Mark Kerr, Courtney Packard, Emily Hatesohl, for her sharp eye and perceptive reading, Andrew Yoder and behind-the-scenes folks at Rowman & Littlefield I've not yet met. And I owe a great debt to many already named plus Mary Braun, Elizabeth Swayze, Jackie Marlette, Erik Schlenker-Goodrich, my writing groups, members of the writing community, and friends in environmental groups for continued moral support and encouragement.

There would have been no book without the many people I met and corresponded with—zealous, generous souls who welcomed me into the driving passions of their work. Susan Jane Brown has given me hours of her busy life, sharing background, explaining legal history and concepts, recommending other actors and publications. For "From Conflict to Collaboration," I am also deeply grateful to Mike Billman, Boyd Britton, Karen Coulter, Maia Enzer, Pam Hardy, Andy Kerr, Tim Lillebo (though I wasn't privileged to meet him), John Shelk, and Mark Webb.

Kyle Tisdel and Jim Ramey set me up with history and legal explanations concerning the North Fork Valley's campaign against fracking, and importantly, gave me lists of people to talk with when I visited the valley. I am most grateful to Kyle, Jim, Wink Davis, Shirley Ela, Daniel and Joann Feldman, Brent Helleckson, Alex Johnson, Pete Kolbenschlag, Bob Lario, Susan Raymond, Sarah Sauter, Robin Smith, Tom Stevens, and Mark Waltermire who love the valley, fight for what they love, and gave time from their busy lives to tell me about it. Thanks to Joe and Katherine Colwell who hosted me at their fascinating Colwell Cedars Retreat during my week in the North Fork Valley. Thanks to Yvonne Young who connected me with her aunt, Shirley Ela, and to Shirley for a lovely hour's history and later reading of the story for accuracy. Big thanks and a healthy dose of awe to Natasha Léger who has been hanging with me on the years since my visit and gifting me with encouragement, updates, answers, and introductions. And I send my great admiration for the dauntless work of Citizens for a Healthy Community, Solar Energy International, the Western Slope Conservation Center, and the Farm and Food Alliance. May you never give up or give in.

Thanks to Mary Christina Wood for time and suggestions, and especially for her book, *Nature's Trust*, and her work in atmospheric trust that helped launch Our Children's Trust. Thanks and admiration for their hard work to Julia Olson, Meg Ward, Caitlin Howard, Andrea Rodgers, and the many dedicated attorneys and staff who do the important work of fighting for our children's future. Special thanks for the inspiration and infusion of hope I received from Tyee Maddox Atkin, Ian Curtis, Stella Drapkin, Corina MacWilliams, and Serena Orsinger, and to my son Jeff Hess who arranged my meeting with these remarkable young people.

The many organizations working hard and joining forces to protect land, wildlife, coastal jobs, clean water, private property, and the climate from the construction of the Jordan Cove liquid natural gas export facility and its 229-mile connector pipeline are listed separately, but I am inspired by and grateful for the work of them all. Special individual thanks to Lesley Adams, Kathy Conway, Steve Dieffenbacher, Francis Eatherington, Alex Harris, Robyn Janssen, Alan Journet, Jody McCaffree, Stacey McLaughlin, and Hannah Sohl for information, clarification, and counsel. You folks are doing an exemplary job.

For help and information on the exciting Transition Movement I am particularly grateful to Judy Alexander, Sandy Bishop, Richard Dandridge, Kate Gessert, Don Hall, David McLeod, Rhea Miller, Marissa Mommaerts, Jeremy O'Leary, Madge Strong, and Marissa Zarate. Transition US offers many online classes and has a terrific website with loads of information on this most promising movement.

I am further indebted to all of the people and groups who contributed artwork. Thanks to Dave Imus of Imus Geographics for the good-looking and informative maps and to Oregon Public Broadcasting, BMFP, Mike Billman, Oregon Wild, Ochoco Lumber Company, Smith Fork Ranch, Rita Clagett, WELC, TEDX, Natasha Léger, Robin Loznak/Our Children's Trust, Rogue Riverkeeper, keithhenty. com, Allen Hallmark, Francis Eatherington, Jason S. Squire, and Marissa Mommaerts for beautiful photographs introducing readers to people and places discussed. A special shout-out to Celina Johnson-Hess, my granddaughter, for the author photo.

Finally, now and always, deep love and gratitude to my dear husband, David, our children (who long since stopped being children) Erika and Jeff, their spouses, Bob and Anna, our grandchildren, Nate, Celina (and Geoffrey), Tasha, Camila, Owen, and Margaret, and our great-grandchildren, Evangeline and Benjamin. You all hold me up and keep me going. Thank you!

Prologue

ON DECEMBER 13, 2017, David and I celebrated the twenty-fifth anniversary of our abandoning the mainstream. A quarter-century earlier, we had left our convenient house in Eugene, just a mile from the University of Oregon. We left its plumbing, its electricity, its *normalcy*, for twenty-one wildly wooded acres in the hills—with no house.

This was not a socially or environmentally conscious move; it was a move of economic urgency. Renting out our house in town could help us recover from debts we had incurred in a recent downturn. Maybe it was also a bit of Sagittarian foolhardiness or thrill seeking. So what if there wasn't a house? We enjoyed roughing it, didn't we? But whatever the combination of motivations, we plunged ahead.

And so just over a week before winter solstice 1992, we shoved aside tools and fertilizer in the ancient singlewide trailer that had been our nursery's storage shed, moved in a bed and a two-burner stove, and began our camping adventure. When the weather permitted, we moved to our summer home, a tent by the pond on a lower shelf of our south-facing slope. Thinking back over those twenty-five years on our forested land in Oregon's Coast Range foothills, my memories are mostly happy ones: the surprises, the lessons, the rewards.

In our woods, we listened to owls hoo-hooing through seasons of courting, nesting, and raising hungry chicks. We listened to coyotes yipping and howling, to the warbling of wood warblers as they returned in the spring, and to crikking of crickets in the fall.

Without electricity, we became more aware of and more enamored with natural light, eagerly watching the evenings growing longer in

December, the sun rising ever earlier beginning in January. As earnestly as we cheered for each new moment of light, we were at least equally captivated by the dark. This far out of town, the Milky Way looked truly milky, the stars not dimmed by city lights. We thrilled to meteor showers and eagerly anticipated each season's constellations.

We thrilled as well to other signs of seasons. Over the years, we learned to use the changes in plants and activities of animals, the smells and sounds in the air, and the length of daylight as our calendar.

It was a time of intoxicating revelation, but our ability to focus on the sounds and smells, the growth and movement of seasons, came directly from our distinctly abnormal living situation: no insulation deadening the sound, minimal walls obstructing the view, no blare of television competing for attention. Clearly, there is also a challenging side to thin metal or canvas walls, to being without electricity or indoor plumbing. And those things took some getting used to. But that was mostly a matter of breaking old habits and expectations and finding new ways to do what needed to be done, powerful lessons.

The tent and trailer provided shelter of a sort. Water poured out of a couple thousand feet of poly pipe we laid down the east slope of our hill, and came from a well we shared with a neighbor. The shutoff valve at the end of the pipe was close enough to our trailer that we could carry water for dinner or washing up without too much trouble. To stay warm in frigid weather, we burned propane in a little metal stove that quickly warmed the trailer to a low bake. In lieu of showers, we heated water for sponge baths. We had a pit toilet, and later, buckets of sawdust for composting. We grew abundant produce, and when our schedules or the hungry wildlife didn't allow a good harvest, we subscribed, gratefully, to weekly boxes from a neighboring farm.

It was a surprise to discover how fulfilling it felt to find simple ways to provide for our needs. It felt clever, like a little child might, learning a new skill, to manage without the switches and pipes and buttons we'd long assumed would be forever standing faithfully by. And even more fulfilling was the gratitude we felt for the warmth of the stove or the sun, the clear pure water that we carried, a campsite by the pond. We were grateful *because of the inconvenience* in acquiring these essential things that in our past convenient lives we had taken for granted.

Sometimes that gratitude became overwhelming as our unhoused lives connected us emotionally with millions of other unhoused, so many of them without one or more of the requirements of life: adequate food, water, shelter. As we carried clear water that had come from our well, I thought of women trudging multiple kilometers with jugs of water on their heads, or the toddlers I saw pictured, squeezing their drinking water from rags dipped in mud puddles. As we lit the propane heater in our narrow trailer, I thought of places where ten people would live in a shack no bigger than ours, with body heat or an open flame for warmth. Our lives were challenging but joyous; their lives—some of them—were desperate, bringing neither joy nor hope. Living in the woods, we keenly felt the inequity that had been largely invisible to us in our former ordinary lives.

Though deprivation could be called living simply, simple living need not entail deprivation. In our case, we had the necessities: potable water, land, rudimentary shelter. Doing for ourselves and undistracted by electronic passive entertainment, we found rewards we'd never dreamed of. For decades, we'd been aware that overconsumption, resource exploitation, and the burning of fossil fuels were ruining the planet. But if simple living means discomfort and anxiety, how many are going to sign up? I found it beyond gratifying to see for myself that such a life was not only possible, but could be enriching. At the same time, I anguished over the predicament of all those millions of humans living desperately and suffering.

My book *To the Woods* is a story of those first fifteen years. *Building a Better Nest* begins when, after sixteen years of "camping," we were able to begin building a real house. Wanting to honor all of the things we had learned, we retained and enjoyed many "simple-living" ways (off-grid solar power, rainwater collection, use of gray water), but we began to realize that even as personally rewarding as it was, our small house and its eight photovoltaic panels were pretty insignificant in planetary terms. We would do precious little to slow climate change and nothing to help those who were truly suffering. At the same time, the mess in the world became daily more urgent: food insecurity, poverty, economic inequality, forced migrations, and the signs and terrifying projections of climate change. Besides the growing challenge to the planet, the foundations of democracy seemed under attack in our nation as in many other countries. How can a person respond to so many assaults?

It became clear that each item on that list of world messes is exacerbated by climate change, which itself is largely caused by the acceptance of social inequities. So specific climate issues and the crux of the inequities must both be tackled, demanding swift and significant change. The most effective avenue for doing that is in joining others, exercising people power, the way that major social change has always been accomplished. It will require people finding others with shared goals and interests, people learning to listen to those with divergent needs or backgrounds, people with the energy and courage to stand up for the needs of other people, animals, and ecosystems—even when that necessitates challenging the powerful. It will require people working together for the planet, for democracy, for each other.

Introduction

P AUL HAWKEN'S *Blessed Unrest* was published in 2007, the year before David and I began building our house. As I was ruminating about the necessity of bottom-up group action to accomplish necessary change, I was delighted to learn about the health, energy, and diversity of groups that Hawken described, groups working in service of both human rights and the environment. Worldwide, he counted at least 30,000 environmental organizations. An Indigenous person reminded him that human rights and environmental rights are inseparable, so, adding human rights groups, he calculated that the number topped 100,000. As he continued probing, he decided those estimates were far too low. He projected there were a good million, and maybe as many as two million, such organizations throughout the globe.

I believe groups like these represent our hope for the future. Though policy change is clearly essential, no charismatic leader or altruistic party will solve the world's problems. If the problems are to be solved, it will be up to the people—people who can be inclusive, who can see the big picture, who will fight for both social and planetary needs. The world needs people who prioritize health of communities and the planet over wealth accumulation for the few and then join with others to compel our leaders to initiate effective laws and practices to that end.

Day to day, I fret. I march. I reuse and recycle. I write letters and vote. I live extremely simply. But all of that is miserably inadequate. I have grandchildren whose mature lives may be spent on a planet my parents would not recognize. I have great-grandchildren whose home might be such a netherworld for most of their lives. Eleanor Roosevelt asserted, "A democratic government represents the sum total of the courage and the integrity of its individuals" (*Tomorrow Is Now*, 1963). So how best, I wondered, can *this* individual serve?

In *Blessed Unrest*, Hawken called the tremendous surge of people-power the largest movement in the world that no one had noticed. A movement gains power when others see it happening—others who might support or join or emulate or be reassured that they're not alone in their own pursuits. In the decade since Hawken's book was published, that surge of energy has not diminished, but the world is still not generally aware of the work being done on earth's behalf by dedicated people, aware that there is a path to hope, a path welcoming us to join. Perhaps, I thought, I can do some good by spreading the word.

So I went searching for people working together on environmental problems, the environment being the womb of all life. When I was working at the University of Oregon, I admired an environmental clinic the law school had set up, where law students could provide legal representation to grassroots conservation organizations across the West, drawing on the support of fundamental US environmental laws.

In 1987, the year before I left my university job, clinic law students challenged timber sales that threatened ancient forests and the wildlife, including the endangered spotted owl, that requires those forests as habitat. This set up a kerfuffle between the timber industry and the university. To end the dispute, in 1993 the clinic moved off campus to become a separate nonprofit public-interest organization named the Western Environmental Law Center, or WELC. Public-interest law focuses on the concerns of underrepresented people or causes, giving voice to the voiceless. Work is done pro bono, the organization relying on charitable gifts from individuals or foundations to support its mission of protecting public lands, wildlife, and communities.

In the last three decades, WELC has spread its reach, now having offices in Washington, New Mexico, and Montana, as well as in Oregon. It reports working with upward of two hundred partners and allies annually on environmental and social justice issues. WELC would seem an obvious organization to ask for what I was seeking: information about people working together for the environment and each other. So I trekked over to the WELC office in search of exemplary projects.

❧

Hillary Johnson, then Communications Director at WELC, sorted through shelves of files, suggesting a handful with particularly interesting stories and strong people-power action. I selected four: two in Oregon, one in Colorado, and one centered in Oregon but active throughout the nation and abroad. To those, I added a group that is focusing on ways of living that promote equality, diversity, resource protection, community resilience, and environmental stewardship.

I was excited to learn that even with the nation becoming ever more fractious, there were groups of former enemies who had learned how to listen and talk to each other for mutual benefit. Listening and learning to trust "the other" seems fundamental for any cooperative venture to be effective. A beautiful example was eastern Oregon's Blue Mountains Forest Partners, a seemingly unlikely collaboration of conservationists and timber interests.

Hillary introduced me to WELC staff attorney and Public Lands Co-ordinator Susan Jane Brown, who told me how the BMFP collaboration had come to be. David and I drove across Oregon's Cascade Mountains to the town of John Day to meet other players and try to understand how those groups established trust. Their story, "From Conflict to Col-laboration," is the first in the book.

I boarded Amtrak to explore the ways a scattered rural community on Colorado's Western Slope of the Rockies stood up to the oil and gas in-dustry. Hillary had told me that Colorado's densest concentration of or-ganic farms, orchards, and vineyards was found in this valley. I was eager to see the valley firsthand and to meet people who lived there. WELC Energy and Community Coordinator, attorney Kyle Tisdel, helped me sort out the history and legal aspects of the community's struggles and shared an invaluable list of contacts. In the North Fork Valley, I visited with organization leaders, activists, farmers, ranchers, vintners, and artists. This valley's story is number two, called "Community."

In story three, wise and eloquent children in my hometown of Eu-gene, Oregon, as well as in states throughout the nation and the world, fight for their right to a livable future. I explored legal frameworks and concepts: What was the atmospheric trust doctrine and how did it get started? How do the wording and interpretation of environmental laws play out in climate considerations? What are the slow but essential steps in the judicial process? What grassroots youth groups spun off from the kids' court case? This story, "Borrowed from Our Children," left me optimis-tic about the upcoming generation's ethics, courage, and determination.

I marveled at the many groups—those protecting the ocean, the rivers, the Klamath-Siskiyou Mountains, the climate, their homes—that joined forces with native tribes and landowners in southern Oregon to fight a liquefied natural gas export terminal with the potential to threaten land, water, sacred Indigenous sites, and the climate. This is story four, "Converging on the Cove."

In those last three stories (stories two, three, and four), like in a bucket brigade against a house fire, diverse people work together against a common foe. In each of those middle three, we see that the adversary is not so much a particular person or organization as it is a philosophy of life: a basic paradigm focused on the primacy of short-term profits over long-term health of the biosphere and its inhabitants. These are not quickly resolved conflicts. Though I wish it were otherwise, I don't expect a "happily ever after" in my lifetime. But the organizations and the energy, intelligence, and dedication of these people and groups give me tremendous hope.

Story five, "Getting It Together, Together," describes a growing movement that's finding new ways of living in ecological harmony rather than dominance. Tackling a full spectrum of problems—from economic inequity to food insecurity, to soil, air, and water degradation, and more—they find a model in healthy, functioning ecosystems. Understanding the essential nature of empathy and benefits of diversity discovered in "From Conflict to Collaboration," these "transitioners" also see the necessity for the paradigm shift confronted in the middle three stories. Notably, they recognize the overarching importance of one's own inner health as well—the health of the soul, so to speak. They have set their compasses for building an equitable world where the present and future well-being of the biosphere and its species is paramount.

In researching this story, David and I traveled to the far northwest of Washington State to talk with some of the folks involved in transitioning toward community resilience and equity. Across the nation and the globe, many are waking to the realization that resilience is found in diversity and connections, in bottom-up organizing, in local self-sufficiency, in honoring natural systems and all of life on earth. This movement recognizes and addresses the complex interweaving of problems: resource depletion; the causes, damage, and threat of climate change; rampant social inequities; and the sensibilities of the human spirit that can either thwart or facilitate change. Most of all, this movement thinks in terms of "us," not of "me," and believes that by working together at a local scale, we can accomplish sustainability and resilience probably impossible when dependent on outside contribution.

My travels, interviews, and research have been a heartening exploration. In spite of the daily chaos and dissension on the news, I find real promise in the numbers of people who join together to do the right thing. The more I look, the more I find people motivated not by personal profit or acclaim, but by fighting for or living in ways responsible to the planet and to one another.

Their work and dedication are both exciting and inspiring. Potentially, some could save ecosystems that support humanity, addressing jobs *and* the environment. Conceivably, their efforts may rescue our threatened democracy. Unquestionably, working together will boost a personal sense of purpose and social camaraderie. But such work is not easy. It takes empathy and understanding; it takes determination and patience; it takes focus and courage and more than a little love.

From Conflict to Collaboration   1

SUSAN JANE BROWN SPED DOWN the Portland State University sidewalk, checking building names as she ran.

Ah! There it is. Now, which hall? Which room?

It was early 2003. A meeting for the already contentious Forest Plan Revision for northeast Oregon's Blue Mountains was about to begin, and Susan Jane was late. At last, there was the room. She dashed in, breathing hard, to standing-room-only. *Sheesh!*

A crowd like this had not been on her radar. Usually, she could expect a few Forest Service officials, perhaps a representative or two from the environmental community, maybe a delegate or spokesperson from the geographic area being discussed. Who were all of these people?

The petite redhead stood on tiptoes trying to find gaps over shoulders and between heads. Her heart did a little somersault seeing near the front of the room a pair of men she remembered well. *The timber beasts are here!*

The last time she had seen those two, she was cross-examining them in Grant County Court. The case had gone well for her clients. They had successfully shut down two ill-conceived timber sales, including some for "salvage-logging." *Salvage-logging! Saving logs? More like destroying habitat and the forest ecosystem.* Fortunately, the judge got the message of the importance of downed logs and snags as wildlife homes, moisture reservoirs, and barriers to erosion. Susan Jane's clients won their case, but few in the logging community were sympathetic with that decision.

Susan Jane peered toward the front of the room, trying to focus on the speaker. Near her two forestry acquaintances, she saw their backup. Maybe a dozen big men in flannel work shirts, jeans with suspenders or wide belts, work boots—the logger uniform—and others who were likely owners or managers. No surprise that they were here, of course. The Blue Mountains in question were in their backyard. By this time, she had been spotted. Heads turned her direction; folks whispered, gestured. Susan Jane was hugely outnumbered. But then, she usually was.

At a break between talks, Susan Jane left the room in hopes of finding a colleague or two, and perhaps even snagging a seat closer to the front for the next talk. As she threaded through the crowd, suddenly her way was blocked by a compact fellow with graying hair and perfect finger-curls on the ends of a narrow handlebar mustache, his dark eyes dancing under a flowered baseball hat.

"You're Susan Jane Brown, aren't you? The little gal who's been whipping our butts in court?" This was Boyd Britton, Grant County Commissioner.

"I am," Susan Jane said, pulling herself up to her full five feet, two and a quarter inches.

Behind the commissioner, three burly timber dudes sauntered up. Susan Jane checked over her shoulder, looking for allies. But none of her friends were near, only six-foot guys in logger getup. Some of them looked familiar. Perhaps they had been drivers in the convoy of big rigs that always tailgated her after administrative appeal meetings, ushering her to the Grant County line on her way back to Portland.

§

The commissioner and his timber and ranching constituents had arrived for the forest-planning meeting in Portland thoroughly pissed off. The very symbolism of the day was too much. Grant County had just sent *two busloads* nearly *three hundred miles* to have a bunch of white-collar so-called experts tell them how to manage their own forests. These folks wouldn't live in the high country of eastern Oregon, surrounded by fossil-rich painted hills and forested Blue Mountains, if they didn't love it there. They and their parents and their parents' parents had all worked these forests and didn't need a bunch of citified flatlanders thinking they knew better than the folks who'd spent their lives there. Besides, what did they understand about the jobs the trees provide? Grant County needed the work. It was the mills that kept the communities alive. Without logs, John Day would wither up and die, as would Prairie City and Canyon City and most of the rest of the county. But for years now, timber sales had been tied up in court, and logging had come nearly to a standstill. *That little Susan Jane Brown is a smart one, no doubt about it, but does she care as much about people as she cares about trees?*

Standing in the hallway between talks, Susan Jane looked Commissioner Britton squarely in the eye and averred that yes indeed, she was the attorney who had been successfully challenging their local timber sales. The hairs on the back of her neck bristled as a coterie of the logging community coalesced from all sides.

"Well," Commissioner Britton said, "you're doing a good job. But it's not working for us. I'd like to invite you to join some of my friends and me in a walk through our woods. I'll give you a ride over. And I'll even bring you back."

Susan Jane searched his face. He had a straightforward, honest look. He didn't appear to take himself too seriously, although she was sure he was quite serious about the cause. She considered his invitation. It wasn't a matter of trust, really, but being open to possibilities. She would hear him out.

"Okay," she said. "I'm in. I can't pass up a free 550-mile round-trip chauffeur."

"The Commish" Commissioner Boyd Britton.
Photo by Amanda Preacher, courtesy of Oregon Public Broadcasting.

The Making of an Activist

The fact that Susan Jane was contemplating trekking through the woods with the very folks she had been opposing in court was but one more side branch in the multistrand river of her life. At one time, her life's strategy had seemed straightforward—a clearly channeled riverbed. From elementary-school days in Colorado Springs, she had had a plan. Growing up with parents enmeshed in keeping their business in the black, she had from her earliest memory been aware that profit versus loss, spend versus save, were drivers of her family's attitudes and decisions. She determined that money concerns would never constrain the shape of her own life. She would be wealthy: she would be a lawyer.

With that in mind, the path was clear. She worked hard in school, knowing good schools required good grades. She was on the debate team all four years of high school in Valparaiso, Indiana, and continued in college. An excellent student, she was accepted at prestigious Vanderbilt University, in Nashville, Tennessee. There she majored in political science and graduated *cum laude*.

When not busy at school or school-related activities during her growing-up years, Susan Jane was outside exploring, capturing worms and frogs, playing in the creek; or later, swimming, skiing, or at summer camp. She came to connect her school's environmental debate themes to the natural world and to governmental policies. Environmental law, she decided, would be the perfect specialty.

So in the late 1990s, Susan Jane applied to Lewis and Clark College in Portland, Oregon, the number one environmental law program in the country, from which, according to her childhood plan, she would go to the pinnacles of wealth and fame and would take her place among the movers and shakers of her generation. But as is often the case in the life of a twenty-one-year-old, her goal-bubble was about to bounce over the sharp spikes of reality.

Susan Jane had learned early in her conservative, middle-class neighborhoods east of the Rockies that polite society had rules governing manners and suitable apparel. Her grandfather was a successful and well-respected businessman, and Susan Jane knew how to dress and behave with his friends, just as she did dancing the cotillion with friends of her own. At Vanderbilt, Susan Jane wore skirts and heels, the appropriate classroom attire. So it was a shock when this proper young lady stepped onto the campus of Lewis and Clark to see, on this sunny August day, not the suitably dressed, well-bred students she was used to, but what appeared to be made-for-TV hippies and hoodlums. Certainly these barefoot and dreadlocked creatures weren't students. Bare feet might be acceptable at the beach, but school—especially law school—was serious. If *they* weren't out of place here, maybe *she* was. Here in the Pacific Northwest, she didn't know the protocol.

That cultural shock was soon followed by the reality of first-year law school. Nothing fun. Just boring basics. Nothing that would teach her about this western environment, so foreign to her. So, a self-starter since her early days as a latchkey child, Susan Jane went in search of her own stimulation. In 1969, a group of Lewis and Clark law professors, students, and alumni had established the Northwest Environmental Defense Center (NEDC) to work for the protection of Pacific Northwest natural resources and environment. Becoming part of the NEDC, Susan Jane joined dedicated students who worked with environmental nonprofit organizations on lawsuits. It was an exciting time for environmental law. The early years of the Northwest Forest Plan, instituted in 1994, and waning days of the Salvage Rider, a short-term provision exempting salvage timber sales from challenge under environmental laws, concluding in 1996, gave the environment a few laws on its side for the first time in a long while. Susan Jane learned about controversies in the forests, wrote arguments, immersed herself in the action.

The NEDC worked and explored in the field as well, and when some older students announced a weekend field trip to the Willamette National Forest's Detroit Ranger District, she had the perfect opportunity to see up close what she had been reading and writing about. She did have a moment's hesitation when the others arrived with all their gear the morning of the outing: she was the lone female. *What would her soon-to-be fiancé think about her backpacking with a bunch of guys?* But her ambivalence was brief.

It turned out to be a mind-blowing forty-eight hours. It was the very first time she had set eyes on a Western Cascades forest, the first time she had seen a rainforest, the first time she had absorbed the magnificence of old-growth trees. Before that weekend, she had no idea that trees could grow so big or that forests could be so beautiful. It was also the first time she had to look at the littered, barren, stump-filled wasteland of a clear-cut. And the first time she realized that rather than being the home of a benevolent Ranger Rick, surrounded by soft music, flowers, and baby animals, the job of the Forest Service was to sell logs. She saw such majestic beauty on one hand—and on the other, exploitation and total desolation.

It was a dramatic, head-spinning awakening. She said she felt as if her brain had exploded. She was both shaken and inspired. She vowed right there that under her watch, not one more old-growth tree would ever be cut. Not only her goals, but she herself was changed for life. Goodbye straight and narrow channel; goodbye rich and famous. Hello public-interest environmental law.

Newly committed to the work of those passionate and energetic folks at the NEDC, Susan Jane began her "forest watch" activities with students and other dedicated environmentalists who monitored timber sales that were under contention. One such "watch" was near Randle, Washington, in the Gifford Pinchot National Forest. The Forest Service was planning to exchange forestland, including two hundred acres of 450-year-old trees, for a larger acreage of cutover land owned by Plum Creek Timber Company. The townspeople of Randle, long a timber town, knew Plum Creek well, having been the unwilling recipients of numerous mudslides caused by the company's previous logging. Now the company would not only be cutting the western-most old-growth in southwestern Washington, they would also log Watch Mountain. Rising 3,500 feet above their community, the mountain would almost assuredly come rushing down on the people and their homes in the first rainstorm once the trees were gone. It had happened before.

So it was here in the forests near Randle that students from Evergreen College in Olympia and fervent environmental activists (who looked just like those students so alarming to Susan Jane when she first came to Lewis and Clark) made platforms in trees and settled in. Soon wives and daughters of loggers began bringing the demonstrators food and drink. The timber industry had provided the livelihood for these women, but now their homes and families were being threatened. They were grateful for the possibility of protection promised by the tree-sitters.

This eye-opening alliance for common goals inspired Susan Jane to join forces with the local timber industry on the Gifford Pinchot Task Force in their work to restore and protect those Cascade communities and ecosystems. Together they would unite diverse interest groups, bring sustainable jobs to the rural areas, and help shape Forest Service policies. That collaborative work undoubtedly helped her to find Commissioner Britton's proposal intriguing. And it was from the Gifford Pinchot collaboration that she would bring a logger friend to accompany her on the proposed walk through the forests of Grant County.

A Walk through the Woods

Susan Jane and her logger friend, along with Commissioner Britton and a dozen or so Grant County lumbermen, stood in the cool shade of the Malheur National Forest. The sun had blazed as they drove through sagebrush, bunchgrass lands, and acres of gnarled juniper trees. Now, beneath towering conifers, the temperature had dropped at least ten degrees.

"What do you see?" asked the commissioner. Around them were lodgepole pines, Douglas firs, delicate deciduous-needled larch trees and true firs. What Susan Jane didn't see were big old "yellow-belly" pines—the giant ponderosas that once were dominant in these forests. Many small trees, trees more fire-prone than thick-barked ponderosas, and snarls of underbrush all were testament to past years of over-harvesting mature trees, recent years of no logging, and the decades-long reign of Smokey Bear. Fire is nature's way of thinning and cleaning up a forest. It was hard to square the Forest Service's fire-phobic management approach to anything that would develop ecosystem health in these dry east-of-the-Cascades Oregon forests.

And what did the commissioner and his colleagues see? They saw potential logs going to waste, potential jobs unrealized. They listened patiently, but overharvesting of ancient trees and controlled absence of fire were in the past and not their fault. They didn't understand why the environmentalists couldn't see that. What could be done now? How could they get their jobs back? They walked and talked for three days and weren't sure they'd made much headway toward mutual understanding. But they agreed to keep talking. And to keep listening.

Managing the Blues

Different perspectives on the purpose and management of the forests were nothing new to the Blue Mountains, a four-thousand-square-mile range lying in the northeastern corner of Oregon and stretching into southeastern Washington. One of the issues was from misunderstanding the soils, so different from soils farther east on the continent. In her book *In Search of Ancient Oregon,* Ellen Morris Bishop (2006) explains that layers of ash produced the soil of the Blue Mountains, where Oregon's oldest history lies.

About 286 million years ago, as earth's landmasses merged to become the continent Pangea, geologic plates began to shift at the edges of the world-ocean, causing volcanoes to erupt on island arcs far from the shores of what would become North America. For another nearly 150 million years, Oregon didn't exist and the land that is currently Idaho hugged the coast, but as the seafloor plates moved and North America shifted west, about 150 to 90 million years ago, some of those island arcs collided with the continent, volcanoes blowing with every collision, forming the genesis of the Blue Mountains and the inception of what would become Oregon.

Sixty-five million years ago, repeated layers of ash and lava poured over the area. Forty million years later, during the Miocene, fluid basalt flowed up through lava holes, layering basalt over the lava and ash. Through thousands of years, as the softer layers eroded more quickly, deepening riverbeds exposed geologic history on canyon walls. About 7,700 years ago, Mount Mazama blew and blew and blew, creating Crater Lake and depositing three to four feet of new ash over the Blue Mountains. These deep-ash soils grew enormous trees—ponderosa pines with trunks five and more feet in diameter—but when compressed by traffic, machines, or grazing, these soils turned hard and lifeless. And when trees or other vegetation disappeared from steep slopes, the soils washed away. These soils, along with the area's climate, portended both promise and caution to management of the Blues (Langston, 1995).

Long the center of trade between native tribes from the Pacific Coast and Plains nations, the Blue Mountains quickly became a focus of struggle once Europeans arrived. First came fur traders and miners, then loggers, farmers, and ranchers. Some sought commodities, some a new home or a new way of life. Homesteading often required development, and settlers seemed to feel that they not only could, but *should* improve on what nature had provided, "improving" usually meaning clearing trees and planting crops or running livestock. Some even felt that the natives had no right to the land because "they didn't work it." It wasn't long before these disparate visions began to conflict with each other and with the land itself.

Letters and diaries of early white emigrants tell of their relief in seeing trees when they first came to the Blues after traveling through miles of sagebrush, and their joy with the open, parklike ponderosa forests, which reminded them of the managed landscapes of home. But they were appalled that the natives would set fire to the beautiful forests. The natives—Walla Walla, Cayuse, Umatilla—who had lived for thousands of years in this semiarid land, had learned the appropriate season to burn different kinds of forest for various management goals. In dense stands of fir on north and east slopes, small fires set after the beginning of the fall rains created openings and edge habitat where game could graze and fruiting shrubs such as huckleberry and grouseberry could grow. Burning when the forest was moist decreased the chance of fire escaping to these more susceptible trees. On the other hand, repeated light burning in fire-tolerant ponderosa pines, more common on sunny south and west exposures, kept the "parklike" openness so admired by early European-Americans by eliminating common understory saplings of more fire-prone Douglas and grand fir along with other woody undergrowth.

The white newcomers, however, saw fire as destructive and worked to exclude it as best they could. Soon the forest character changed: no longer were the pine forests open. Congested undergrowth shaded out pine seedlings, and lightning found ready fuel accumulated on the forest floor, causing fires that were far more destructive than the planned burns of the tribes.

People were in awe of the huge trees but cut them to sell the timber or to clear land for farming. Lush, abundant bunchgrass seemed to call for grazing, but as grazing pressure increased, those ash soils compacted and grass struggled. Tension built between local cattle ranchers and big out-of-state cattle corporations and between cattle ranchers and sheepherders. The sheep also belonged to out-of-state corporations, but usually were handled by nomadic herders, which gave one more reason to resent them: unlike the settlers, the herders had no earned property rights. Stock trampled grass and ate it down to dust. Soil eroded, cheatgrass replaced the lush bunchgrass meadows, and battles broke out between competing interests.

Many homesteaders gave up as their farms failed or livestock died. They moved on, confused, because the luxuriant growth had suggested such promise. But what few could see through all of that green were the constraints of soil and climate. Land that supported healthy forests could not necessarily translate to thriving farms once the trees were gone. The deep but vulnerable ash soils had nothing in common with the soils the homesteaders knew, and nothing in their experience prepared them for the temperature extremes or the paucity of water.

Though the annual precipitation varies widely across the Blues, some southerly areas receive less than ten inches a year, this falling mostly as snow, and those same places can experience temperatures as low as -54 degrees Fahrenheit and as high as 110 degrees Fahrenheit in the shade. North and east slopes are much colder and moister, and therefore hold snow longer than south slopes, but even there, anything decreasing the ability of those ash-based forest soils to store water, such as compaction by animals or machines or removal of vegetation that holds the soil in place, directly affects the health of the entire living community.

Still, whether for clearing land to farm or ranch, cutting trees to build homes, or harvesting timber for sale, the forests were cut at a rate I find quite amazing, remembering that the pioneers were using hand tools. And there was apparently no compunction against clearing forests. A common attitude was, and for many may still be, that forests were a collection of resources—timber, animals, minerals—to be used for the service of civilization. Not using them was considered wasteful, both

BLUE MOUNTAINS, OR/WA

0   50   100 mi

The Blue Mountains, showing John Day, Prairie City, and Canyon City. Also see Prineville, home of John Shelk and Ochoco Lumber, to the west.
Map by Imus Geographics.

economically and morally—an odd concept to those who love a forest for itself, but one more easily understood by thinking of a produce garden with ripe tomatoes, squash, peppers, and melons hanging on the vine, beginning to crack and rot. Most would consider that waste. Though, truth to tell, there too the rotting fruit would contribute to the soil's biological community.

Enter the Forest Service

When timber interests first saw the Blues, 70 percent of the forest was old growth, those huge trees being an exciting challenge to conquer. In her book *Forest Dreams, Forest Nightmares* (1995), Nancy Langston says that by 1895 two hundred loggers were harvesting fifteen to twenty million board feet (mbf) per year in the Upper Grande Ronde watershed—the equivalent of well over a thousand current logging trucks' worth. Logs were pulled to a river on horse-drawn sled and floated to a mill until the steam locomotive allowed forest workers to lay temporary narrow-gauge railway tracks to get their logs to mills.

By around 1900, when a good half of the old growth, along with most of the rest of the accessible timber, had already been cut, federal foresters became alarmed at the rate that forests were disappearing. They calculated that if logging continued unchanged, the forests would last only about another thirty-five years.

Foresters wanted to run the forests scientifically and efficiently for long-term production. They were upset with the careless and wasteful methods of the corporate timber companies, but saw old growth as being equally wasteful. Their goal was to have a steady supply of logs year after year. This necessitated having trees grow as much as possible as quickly as possible. Old growth took up space that could be used by younger, faster-growing trees, so buyers were sought who would quickly log the old trees, particularly the big ponderosas—the "yellow-belly pines"—to make room for a new era of orderly production.

In order to have the efficient plantations they envisioned, foresters felt the need to eliminate not just mature trees, but anything that might compete for space, light, water, or wood production, including snags, downed deadwood, insects, fungi-damaged trees, inferior tree species. Eventually they cleaned up some forests to their specifications, but rather than increasing production as they had imagined, they got insect epidemics, poor growth, and increased fire.

Like the early settlers, they were confused. They wanted to do the right thing, but coming from eastern US forests and schooled either on the East Coast or in Europe, they didn't understand these relatively dry western forests. In the eastern United States, it rains in the summer. These western forests stay dry in the hot weather. While forest litter decomposes in moist forests, dry forests rely on intermittent, usually low-intensity fire to break down forest floor accumulation. But the foresters saw fire as an enemy.

And forestry, like much of the scientific world, was not yet aware of interrelated parts in an ecosystem. Even when their own observations gave them clues, they couldn't take the next step to change policy. Some of the more observant foresters had noticed that pine seedlings fared better nestled near insect-killed, moisture-holding logs, and that squirrels' caches held the thriftiest pine seedlings as well as the ones that over-wintered best. Even so, they got rid of the squirrels because squirrels ate some seeds, they killed insects that might eat wood, and they removed the deadwood that took up space. The only insects they seemed to see were those they thought to be harmful. They saw budworms or tussock moths eating fir needles, but failed to see—or to appreciate if they did see—the hundreds of parasitic wasps, flies, and spiders or the dozens of birds that fed on the destructive moth larvae. In trying to get rid of what they saw as harmful, the foresters destroyed the habitat of the pests' predators, turning a bit of damage into huge epidemics that killed trees and made forests more fire-prone (Langston, 1995).

Still, the Forest Service felt it had science on its side. To save the forests, they must clear them and begin anew. Such an enormous job could be accomplished only by those big out-of-state timber companies foresters had earlier deplored. So sales were given to old enemies and mills were encouraged to increase capacity until forests were cut well beyond a sustainable level. As private forests were depleted, the Forest Service was impelled to increase harvests on public lands in order to keep the mills open. Besides, there was still some old-growth left, getting in the way of well-regulated production. By the 1980s, Grant County timber interests were cutting 200 mbf a year, with hopes of going to 220. Log trucks roared down the streets; mills hummed; cash registers clanged along Main Street's stores. The town was vibrant, and happy loggers filled the bars.

But the environmental community was not so sanguine. Truckload by truckload, they saw the big trees rumbling out of the forests, taking with them shelter and food for myriad species and being replaced by invasive exotic plants and animals and by fish-killing cloudy water from logging's eroded soils. There was little left of private forests. And at this rate, it appeared the national forests would soon be gone as well. Dismayed, they vowed to defend these public lands and their ecosystems, and went to court.

Less than a decade after the 1964 Wilderness Act, Richard Nixon signed into law the Endangered Species Act, and shortly thereafter, the Northern Spotted Owl, which nested in old-growth forests, was declared a threatened species, shutting down logging in huge tracts of land in order to save the imperiled owl's favored haunts. Though the Blue Mountains were beyond the Spotted Owl's range, the controversies about its home environment had made the judicial system, along with the general public, more aware of forests as habitat and ecosystem, rather than merely as a warehouse for timber. So with the help of that insight—plus the 1994 Northwest Forest Plan's requirements to heed the long-term health of the forests, their wildlife, and waterways—environmental attorney Susan Jane Brown was able to make the case on behalf of several conservation groups that over-harvesting forests could threaten the existence of many dependent species: that removing fire- or insect-killed logs and snags deprived the forest not only of food and shelter for wildlife, but of natural post-fire recovery processes and erosion control: and that a healthy forest ecosystem was essential to protect watersheds, clean water, and fish.

The Blue Mountains Biodiversity Project was a particularly active conservation group. Director Karen Coulter monitors the forests, logging thousands of hours in the field and training multitudes of volunteers in forest education, wildlife and plant identification, map and compass work, and the particulars of field-surveying proposed timber sales and grazing allotments. Her group's efforts have stopped new road-building, modified sales on too-steep slopes as well as in roadless areas and in critical wildlife habitats, increased stream buffers, and stopped spraying that would have killed the larvae of native butterflies and moths while it killed the targeted species.

Susan Jane represented the Blue Mountains Biodiversity Project and other conservation groups. With her advocacy, repeated timber sales were stopped, particularly the harmful so-called salvage sales, and logging levels plummeted. Malheur National Forest harvests that had peaked at 210 million board feet before 1994, dropped to 34 mbf by early 2000. Twenty-three eastern Oregon mills, employing nearly 2,000 people, closed. Unemployment in Grant County reached 14 percent.

It was with this backdrop that Commissioner Britton approached Susan Jane at the Blue Mountains Plan Revision meeting in Portland, to invite her to join him for a walk in the woods. Thinking back on it a decade later, Commissioner Britton's strongest memory of those early field trips was seeing his logger friend Charlie O'Rourke walking along, talking quietly with Susan Jane, his arm around her shoulder.

"Oh, I loved Charlie," Susan Jane says. "He was a big bear of a guy—kind, soft-spoken. He took me to see some private jobs he did, and the work was so sensitively done. Selected trees were removed carefully. The ground wasn't torn up. It was the first time I realized that logging didn't have to be ugly."

After three days of forest trips, the group decided another meeting might be warranted. Neither side had changed the other's mind, but each had spoken its piece and heard what the other had to say. Perhaps there was some incipient understanding. Undoubtedly for a few at least, growing respect. They began to realize that all were concerned about health both of the forest and of the economy, though they were far apart

Susan Jane Brown and Charlie O'Rourke, with others, analyzing the forest.
Photo courtesy of Blue Mountains Forest Partners.

on how to accomplish those dual goals. So they met again in a month. And in another month. And another.

As the levels of trust grew, so did the size of the group. And after three years of meetings with community residents, land owners, private forest owners, forest contractors and consultants, members of local and regional conservation groups, loggers, ranchers, federal land managers, and elected officials, they formed a collaborative, the Blue Mountains Forest Partners, complete with bylaws and a board of directors. A facilitator from Sustainable Northwest helped members focus on shared goals, on listening and trying to understand the reasoning in contrary viewpoints, and on treating each other with respect.

Sustainable Northwest is a nonprofit whose goal, stated on its website, is to "bring people, ideas, and innovation together so that nature, local economies, and rural communities can thrive." Early on, as contentious groups come together, Sustainable sets ground rules designed to break down barriers and create trust. People are encouraged to vent their frustrations, using "I" language rather than the accusatory "you." The goal is to understand where a person is coming from, not to refute their statement.

Easier said than done, of course, when for decades the "other side" has been perceived as the enemy. When logs are hard to get in a timber town, small mills and timber workers suffer, and blame the damned environmentalists who closed the forests. Conservation groups see the scarce remaining forests of century-old trees threatened and forests cleared at rates nature can't replace and they direct their ire and fear toward the loggers. But here, the point was not to change minds or argue a particular case. It was to understand, to give everyone a chance to weigh in, and to find common ground.

A big frustration among community members had been the feeling that no one listened to them or cared about their needs. They appreciated that with the Blue Mountains Forest Partners they had a voice where they formerly felt they had none. Once they were able to express their point of view with no retaliation, they were more able to hear the other side.

The collaborative's stated goal was to restore ecological resilience in a socially acceptable manner, while providing economic benefit to the community. For some of the community members, that potential for economic benefit was a reason to hang with the program even if the "green" gobbledygook made no sense at all. But many were still not ready to join the conversation.

Mike Billman

Six-foot and slender, looking sharp in his neatly trimmed dark beard with white sidewalls, Grant County resident Mike Billman walked resolutely toward his office, his eyes downcast, deep in thought. *Another day at work, and still no answers.*

Mike was the log buyer for the only mill remaining in John Day, but where could he find the logs? The supply was so meager, he had to buy a good half of what the mill needed from Idaho, a 250-mile drive each way. How did that pencil out, either in truck expenses or driver time? Yet he had to find logs somewhere. The mill was counting on him, and with the mill, a hundred jobs. *A hundred families. All those children.*

He was concerned about his own job as well. Without his job, would he still be able to live in eastern Oregon? This was where he *chose* to live, where he wanted to raise his young family. Mike was on the school board too, and had been watching school enrollment dipping lower and lower as jobs disappeared.

Billman saw an endlessly growing black cloud hanging over the industry and no hope for the future. The mill would just try to get by, find enough logs to survive month to month. Nowadays, it seemed that environmental interests held all the cards. It was an impossible conundrum. Mike had studied biology and ecology in school and understood some of the arguments of the environmentalists. Still, it seemed to him that trees could be cut on a sustainable basis. Why couldn't there be lumber and jobs and a healthy forest too? But no. The environmentalists had to shut everything down.

Mike was aware that some of the locals—folks from the timber industry and ranchers and other interested people—were having on-going conversations with enviro-types, but he couldn't imagine those talks would go anywhere. He'd been to the Forest Service's so-called stakeholder meetings where everybody sat in rows staring at the front of the room and listened to the person in charge telling them how things were going to be. Then they'd open the floor and let the "stakeholders" shout at each other for a while. He left the meetings with a headache, and stopped going after a few wasted evenings.

Still, this new group was continuing to get together. It must have been a good six months since they'd become official, and they hadn't given up yet. Maybe somehow they were different. That didn't seem likely, but perhaps he should check them out anyway.

Venturing In

On a Thursday evening in the fall of 2006, still filled with doubt, Mike Billman stood outside the restaurant meeting room. But even with so little hope, he ventured inside to look over this new Blue Mountains Forest Partners group.

A woman's voice rang through the murmur of conversations. Maia Enzer of Sustainable Northwest, a mane of chocolate-colored curls breaking over her shoulders, greeted the gathering. After some introductory words, she asked the participants to divide into smaller groups for discussion, gently reminding them that everyone should have a chance to be heard. Rather than the hierarchical and adversarial seating arrangement Mike had expected, people sat at round tables where you couldn't avoid the eyes or facial expressions of the person across the table, whatever you thought of them or their ideas.

A chair was empty across from a guy Mike knew by face and local gossip, but had never met. In every detail the mountain man—full graying beard, a ruff of gray curls beneath his signature bashed-in brown felt hat, plaid shirt casually topped with a vest—this was Tim Lillebo, a man with quite a reputation in these parts. Though a son of Grant County, a former timber faller, and a grandson of a logger, Tim was now on the other side. He seemed to be involved in every argument to do with the forest, and his side was winning. Folks who knew him seemed to like him anyway, but others saw him as a turncoat. If Mike was going to hang out with a foe, it might as well be with a famous one. He took the chair.

People at each table were asked to introduce themselves and explain their hopes for the forest. Tim was first up at their table. His voice soft, but with a warmth and energy that carried throughout the room, he spoke of his love for the forest, his empathy with the needs of the loggers and millworkers, his fervent desire to return the forests to a healthy resilience and to work toward harmony among all of those who cared about the Blue Mountains.

Mike could hardly believe his ears. It was as if this supposed adversary was reading from Mike's script. When it was his turn, he said, "I'm Mike Billman and—what he said," gesturing toward Tim. Tim might be representing "the other side," but clearly, Mike thought, he was no enemy.

Tim was in a unique position. Having worked in the woods, both felling trees and building roads, and being the devoted grandson of his logger grandfather, he understood the timber culture. But his career did an about-face one day when his grandfather told him, "Tim, as we cut, we thought the old growth would never end. Well, it's mostly gone now. You should cherish what little is left of the old trees." And from that moment, saving the old yellow-bellies became Tim's passion.

In school, he studied the science of forest ecosystems and was able to back his zeal with facts, but as new information surfaced, he also had the ability to change course. In the mid-seventies, he put together an eclectic group of Grant County conservationists, themselves an endangered species, and together they went rafting, hunting, fishing, and exploring. They hiked and camped and read Ed Abbey. Agitating against extraction and development, his group was not popular in the area. Some members even received death threats, but that neither deterred nor embittered them. Lillebo knew that in order to preserve the old forests and restore years of damage, he needed the support of people who depended on the land and waters—the hunters, ranchers, fishers, and recreationists. His experience and knowledge of the mountains opened doors, as he was able to talk in the language of people's interests and let them know that he was listening as well. He felt that the only reasonable action when adversaries came together was to stop and talk. If it worked for the other party involved, he chose to do that talking outside, maybe with a fishing pole in one hand and what he referred to as an "adult beverage" in the other. He may not have begun the collaboration movement, but he was an uncommonly well-qualified member.

Tim, always searching for a way to find common ground, keyed in instinctively to interested industry folks, so he quickly connected with Mike. Mike was immediately taken with how intently Tim listened. It was clear that he cared what Mike thought and felt. Tim seemed almost to be looking inside him.

Mike was captivated. The two men discovered they had much in common, and in time became close friends, first getting together at the corner pub for conversations and eventually for annual campouts along the middle fork of the John Day River, in forests of the old yellow-belly pines Tim loved so well.

But now and in months to follow, they sat across the table listening, while some people tried to find common cause and others wailed in dissent. The classical industry position was that the sole purpose of a federal forest was to provide timber. The only meaningful measure of the health of a forest was the annual board feet harvested. *"We've got to get out the cut."*

When conservationists brought up ecological arguments, timber folks countered with economic ones:

The old trees are essential for forest structure and for large cavity-nesting birds and mammals, like pileated woodpeckers and American martens, conservationists argued. *Besides, large wood stores carbon to ameliorate climate change, and big logs are critical for nutrient recycling as well as holding moisture, thereby decreasing fire danger.*

Timber interests challenged, *How can you waste perfectly good trees? Those big trees are worth something now, but in another ten years, they'll start decaying from the inside out. Then the bugs will come and infect other trees. Cut it down now and save its value, plus give room for young trees to grow.*

Do you know how many species use that old log for food or shelter? Try fifty-five kinds of birds, plus amphibians, invertebrates, mammals, reptiles . . . environmentalists pled.

I'm more interested in food and shelter for my wife and kids and for my neighbors and their families than for your birds and worms, was timber's quick retort.

Disparate Visions

When I talked with Mike, he described the crux of the argument as an incompatibility of viewpoints: Silviculture prescribes growing trees like a crop, timing planting and harvest so that product will always be available, while forest ecology looks at variation in age and species of trees, along with their plant and animal communities and the forest effects on the watershed and greater ecosystem. The latter approach is an inefficient way to harvest a crop. The former sees only the trees, at the expense of the forest.

Nonetheless, there were many in the group who tried to hear, to understand, to find a way forward to identify those values that both sides agreed on. Some of the attendees who chose to listen rather than to battle the other side found themselves painted as traitors by members of their group who stayed home. Among some, loggers have a macho reputation, expected to fight, not to turn the other cheek. Only a wuss consorts with the enemy, and to compromise is to surrender. Likewise, some environmentalists expect their colleagues to hold fast to their values, not to play footsie with industry. That's selling out.

Certainly a part of the difficulty in getting people to the table is worrying that the other side will reject their overtures—that they'll be ridiculed or scorned. But even more worrisome might be the fear that their own clan will reject them. From prehistory, acceptance by the tribe has been essential for survival. If a person is cast out, the hyenas, leopards, or alligators may find an easy meal in that exile.

Still, through it all, a core group of Blue Mountain Forest Partners dared share hope that they could find their way to health for both the forest and its dependent human community. Mike Billman remembers that in the beginning, you could have drawn a line down the center of the room as people clustered with their own, often shooting dirty looks across the space to the other side. But as respect and relationships grew, when you looked around, you might have had trouble guessing the ideologies of the people present.

Before long, the mediator came up with the idea of having Mike and Tim act as co-chairs. Both of them wanted to break barriers and have the group be productive. The combination and their example put more people at ease and encouraged cooperation.

It takes time and hard work to build relationships, Susan Jane says. Unlike the work it takes to understand different points of view, it's easy to file a lawsuit, giving you winners and losers. "Litigation is a great tool for stopping bad things; it's not so good at driving good things

Mike Billman and Tim Lillebo sharing an adult beverage—Tim's favorite, George Dickel. Photo courtesy of Mike Billman.

forward," she points out. Relationships require listening with empathy and without judgment. They require knowing what you are talking about—having a good factual basis—and being able to explain it clearly and not defensively. And they need something all parties can agree on. In this case, the group relied on current forest science and the belief for some, the hope for others, that a healthy and resilient community could come from having a healthy, resilient forest.

To fulfill their goal for social acceptance, the Blue Mountains Forest Partners understood that the public needed to be well informed and have the opportunity to take part in decision-making. To this end, the community was invited to monthly meetings as well as to learning forums and field trips where forestry and research scientists shared the most recent findings on forest management. For the goal of ecological resilience, the Partners needed to implement restoration methods that would counteract a history of fire suppression, protect lives and property from catastrophic wildfires, and return the forests to historic natural communities dominated by fire-tolerant species. After discussion, research, and consultation with scientists, the collaborative proposed to the Forest Service certain restoration treatments that would work toward BMFP goals, and at the same time be unlikely to initiate litigation. They would thin out small fire-prone trees and underbrush and conduct low-intensity controlled burns, giving young ponderosas the room and light they require, reflecting the precontact Indigenous management. As of 2018, no timber sales had landed in court since 2003.

Though the conservation community is generally opposed to salvage logging, members of the collaborative were able to identify logs and standing dead trees from a recent major fire that were accessible from an existing road and could be considered hazards. These, along with small timber from thinning projects, were sufficient to maintain the mill for the time being. The timber industry adapted to handle small-diameter wood for heat boiler systems. Now four boilers generate heat for local schools, hospitals, and the airport. From the beginnings in 2006, project sizes have increased from 7,000 to 42,000 acres of land where fire suppression had allowed forests to fill with more flammable trees, putting ever-increasing numbers of people back to work.

Committed to ongoing adaptive management, the Partners began and will continue to maintain a multiparty monitoring program together with the Malheur National Forest staff to be sure that the projects have the desired effects on the land and local economy. They intend that management decisions be flexible and responsive in order to further their ecological, social, and economic goals. If they see that a treatment isn't working, they will change it.

When I asked Commissioner Britton if the timber industry was feeling better about the conservationists now, he said, "They're happy to be working." Certainly some minds were changed, or at least expanded. Some new friendships were formed. But many—perhaps most—continued to look at the forest through different lenses. Blue Mountains Biodiversity Project's director Karen Coulter commented that though she was glad that the collaboration encouraged community dialogue and helped break down polarization, she was uncomfortable with the scale and pace of logging. She says that it is good to work for common ground, but some values cannot be met: a tremendous divide remains between biocentric and anthropocentric values, or as Mike Billman says, between ecological and silviculture specifications.

Still, there were clear advantages in the way things were going, advantages that could be seen through both sets of lenses. Community members who had never felt they had a voice, now had a place at the table to help protect not only their jobs, but also the forests they loved. People were going back to work. The several small towns scattered near the national forests were better protected from devastating wildfire. Working together for common values such as forest resilience, community self-sufficiency, strong salmon runs, wildfire preparedness, clean water, and carbon storage, folks were beginning to understand that with cooperation they could make a difference in the health of this land they loved, while making themselves and their human connections healthier as well. With relationships generally harmonious, attitudes began to relax and some dared optimism for the future. Skies were not exactly rosy, but few people noticed the thunderclouds gathering on the horizons of their lives.

Weather Change

One summer evening in 2012, Susan Jane Brown left the first hot shower she had had in two weeks, thoroughly blissed out from a camping trip in Yosemite. She was startled back into the real world by frantic messages on her voice mail. The president of Ochoco Lumber Company, the parent company of John Day's Malheur lumber mill, had announced at the end of a Friday workday that managing partner John Shelk was going to close this, Grant County's last mill. But the mill was not only essential for local jobs, it was necessary to proceed with forest restoration.

It wasn't that Shelk *wanted* to close the mill, he told me later, having graciously invited David and me for a visit in his spacious office in Prineville, Oregon. For the past two years, he had done his best to keep the mill open for the sake of the people working there. But how long can you go on losing money? Still not recovered from the recent recession, the market was soft. Big companies had continued exporting whole logs from private lands instead of sending them to mills, cutting millworker jobs and stressing many small mills to the breaking point. In a June 2011 *Oregon Live* article, Roy Keene noted that at least five hundred jobs left the Pacific Northwest every week as boatloads of raw logs were shipped to Asian mills. Keene estimated that at that accelerated export rate, 26,400 family wage jobs would be sent to Asia by the end of the year.

Shelk said that private forests, from which whole logs can legally be exported, had been seriously overcut. Smaller mills didn't have to compete with Asian mills for federal logs, which cannot be shipped, and national forests cover large swaths of eastern Oregon, so those were the primary timber source east of the Cascades. However, those trees were generally unavailable. It had come to the point that it just wasn't economically sustainable to keep the mill running.

It felt a bit like déjà vu to John Shelk. In 2001, he had had to close the Prineville mill, the first one his family, along with four other partners, had built back in 1938. The group's mills had grown from that first one to five, at one point employing about four hundred people. Now they had only the one in John Day. The big difference now, Shelk said, was that when he had to close the Prineville mill, he had been quite bitter towards the Forest Service and the environmentalists, whom he felt had not held up their part of an agreement from ten years earlier. Then he had retooled his mill for small trees that the Forest Service said would come, but nearly all of the sales had been appealed, and he was left with expensive new equipment and no logs. The Ochoco National Forest sales dropped from 115 million board feet in 1991 to 6.4 mbf in 2000. The closure of the mill reverberated through the area: there were the 166 loggers; the 950 people manufacturing doors, windows, and moldings; the eighteen-mile Prineville railroad that had run to its connection with Burlington Northern-Santa Fe in Redmond since 1917. The railroad contributed about a million dollars a year to Prineville's economy, and the mill comprised 40 percent of its business. And then there were all the ways the Shelk family gave back to the community—in scholarships and contributions to cultural events and to facilities such as the High Desert Museum in Bend.

A lot had changed since 2001. At that time, the timber community had been at loggerheads with the environmental community for decades. Considered "ecoterrorists" by many in the industry, the environmentalists' goal, it seemed to timber interests, was to completely shut down all timber operations throughout central and eastern Oregon.

That impression had a lot of basis in fact. Tim Lillebo acknowledged that at one time his mantra, like that of many dedicated conservationists, was, "Stop. Don't cut one more tree. Not one more stick." The Wilderness Act defined wilderness as "an area where the earth and its community of life are untrammeled by man," and that was exactly what many were trying to preserve. However, as the timber wars of the late eighties and nineties cooled, some in the environmental community, like Tim, realized that limited logging could help correct unhealthy, dense, small growth stemming from fire prevention as well as from logging that selectively cut big trees. But it was the passionate, uncompromising, wilderness-protecting Tim that John Shelk knew about, surely one of the worst of the crazy environmentalists.

Shelk had heard the rumor that you could actually work with a few enviros in the area, though he was more than a little dubious. Even so, it was with great puzzlement that he discovered that his log buyer from the John Day mill, Mike Billman, had become friendly with Lillebo. Still, Shelk knew Mike to be generous in his relationships and able to find the valuable nugget in a person, so because of his respect for Mike, he tried to soften his attitude toward Tim. Then Tim and Mike worked with environmental groups who conceded that the mill could have some badly needed, fire-killed logs that were near roadways and were considered hazards, with the promise that the loggers would not enter roadless areas. In turn, the groups agreed not to sue to stop the sale. Shelk's trust toward Tim Lillebo began to grow.

Tim recognized a reasonable mind in John Shelk, and around 2006 or 2007 suggested that the three of them, John, Mike, and Tim, get together with Tim's conservationist friend Andy Kerr. Andy had connections with Oregon's congressional delegation, dating from 1979 when he had helped Ron Wyden defeat the incumbent representative in Oregon's third district. Perhaps something could be done in Washington, DC—some kind of legislation that could address the problems of unhealthy eastern Oregon forests and their dependent rural communities.

Shelk was asked, "Do you know Andy Kerr?"

"Oh my, yes!"

"Maybe we could get together with him to talk about the management of public lands."

"I'd rather have a root canal!" Shelk replied.

John told me that from his perspective at that time, if Tim *wasn't* the devil incarnate, surely Andy was that and worse. From temples to gut, John's being cried out against such a meeting. Why would he want to spend time with people who seemed dedicated to thwarting his every project, tripping his every move? Still, his work was currently at a standstill and something obviously had to change. In thinking it through, Shelk recalled stories of the pioneer lumbermen who, after years of dynamiting each other's incipient railroad tracks in hopes of gaining the transportation advantage for their respective mills, eventually realized the greater advantage of cooperation. Together they would connect a railway from their mills to the main line in Redmond. Both would profit.

So, despite housing a deep bubbling anxiety and a multitude of misgivings, John agreed to go. He and Mike met Tim and Andy at Glaze Meadow, Tim's showcase restoration project in the Deschutes National Forest where the forest abuts Black Butte Ranch. The centerpiece of the project was the saving of an old aspen grove being crowded out by young conifers and underbrush. Aspens grow clonally, trees springing from the roots of other trees, so the entire grove may have been a single organism, many hundreds of years old. With startling white bark, shining leaves dancing in the wind, and glowing golden as summer turns to autumn, the grove was a landmark visible from the nearby highway. Undergrowth encroachment from decades of fire suppression not only threatened the grove itself, but also other plants, animals, insects, and cavity-nesting birds whose lives are enhanced by or dependent upon the aspens.

In response to a recent 90,000-acre wildfire in the area, conservationists had joined with the Forest Service and the timber industry to begin thinning at Glaze Meadow with an eye toward bringing back old forest communities and removing fire-prone undergrowth. Tim was excited about the project, loving an opportunity to show it off. Never confrontational, when he talked with someone who had a different point of view, he would suggest another way to look at things. "Maybe, just maybe, it's like this . . ." he would say. "Or maybe not," his hands flying out. But in the Glaze project, he felt he had a win-win-win situation, promoting forest health, protecting the forest from high-severity wildfires, and at the same time providing jobs in the woods.

At the end of the day, with Mike and Tim hovering over him and Andy, and John cognizant of where he was and with whom he was fraternizing, his head felt close to exploding with a blistering tension headache. Nonetheless, the project was in fact both interesting in its restoration of the aspen grove and instructive, as conservationists joined forces with the timber industry. Glaze Meadow showed that collaboration could work. Perhaps it could be used as a national model.

Andy Kerr, whose Larch Company "represents species that cannot talk and humans not yet born," had the knack for finding common ground to meet a need—in this case, to prevent the sort of fire likely to destroy the forest for many decades. He and John began gathering select environmentalists and industry representatives to work with Senator Wyden and help draft legislation that would take the long view in eastern Oregon forest management, restoring health both to the forest and to the timber industry, which would be doing the restorative work.

The first meeting was as contentious as might be expected. Gathering at Nature Conservancy meeting rooms, people sat with arms crossed, glaring across the room and making pointed comments at the other side. In time, however, they agreed to look ahead, decide what the goals were, and figure out how best to reach them. Their work culminated a year later in a bill that, though it didn't get enacted, helped lay the foundation for the Forest Service policy to begin morphing from "get out the cut" to "strive for health."

So with this experience over the past several years, in 2012 John Shelk approached his economic reality at the mill in John Day without the old rancor toward environmentalists. This time, he felt sorry rather than bitter. Among the conservation community, he had found sincerity, intelligence, and even friendship. Unfortunately, that wasn't enough to run the mill. He needed logs.

Susan Jane shook away her Yosemite high and did an abrupt change from backpack mode to emergency response. She called Mike Billman for details on the mill-closure announcement. She called Forest Service folks and people from Sustainable Northwest. She called Senator Wyden, because he needed to know what was happening. Then she met with Mike, John, and Bruce Daucsavage, the mill president from the Prineville office, and together they drafted the "Southern Blues Blueprint," detailing the problems for the mill and for the forests, and what was necessary to address those problems.

John Shelk in the pines.
Photo courtesy of Ochoco Lumber Company.

By the following Monday, Senator Wyden had called a meeting at his Portland office to talk about options. Later the same day, Mike got a call from the Forest Service supervisor, asking what the mill needed to stay in business. Thanks to the Southern Blues Blueprint, Mike could list each problem and its required solution. The supervisor told him that the Regional Forester in Portland wanted that information, and said that Senator Wyden was also interested.

Within days, Senator Wyden had gathered a group including the forest supervisor, the deputy super, the regional forester, Mike Billman, and others for a conference call, the first meeting of an effort that would last a year and culminate by formulating a plan. The Forest Service was to increase the scale and pace of restoration work to double in two years and triple by the end of 2015. They received federal money for additional staff and the NEPA (National Environmental Policy Act) analysis, and the state found funds for sale preparation and for eastside collaborative forest restoration. The work of the collaborative, the Forest Service, private foresters, government representatives, and conservationists produced a ten-year stewardship plan. It seemed everyone was on board.

With plans for more volume to come, plus the efforts of so many at numerous personal and government levels, John Shelk agreed to keep the mill open. He would have to front it for the better part of another year, and maybe beyond that—but as he told me later, closing at that point was just not an option.

Small mills can't compete with the Weyerhaeusers of the world, so they do well to find a variety of specialty products that the big mills can't afford to provide. White fir is fire-prone and never would have survived a century and a half ago when Native people managed the forests with regular fires, but these fast-growing true firs have profited by fire suppression and now compete with the fire-adapted ponderosas for water and nutrients, and they generously deposit seeds to assure the competition will continue in future generations.

Consequently, to enhance forest resilience, white firs would be cut as would other species of fire-susceptible trees that compete with ponderosas, but for the thinning to be economically feasible required a market for the logs. In a small mill in Prineville, Shelk has used white fir to make furniture-grade molding for doors and windows, or, extra-dried in a kiln until it has no odor, to sell for Japanese packing crates where it wouldn't lend unwanted flavor to the product being packed. Characteristics of the also fire-prone eastside Douglas fir differ from the faster-growing Douglas fir west of the Cascade Mountains, making it ideal for *gem ban*, framing for Japanese houses. The straight and tight grain of larch is excellent for laminating stock—the tension and compression members of glued laminated (glulam) timbers. Once machinery is properly adapted, new markets await.

Malheur Lumber, the John Day mill, would open a night shift in the summer of 2014 for small dimension and specialty work. And John Shelk would continue to get together with Andy Kerr occasionally for fellowship and to take the pulse of the forest and of those who care about it. Andy's a smart guy, John says, and a joy to spend time with.

A decade earlier, Susan Jane could never have imagined that she would one day be involved in saving a mill. But that was before she met Commissioner Britton and Charlie O'Rourke, the people of Grant County and the dry eastside forests. Charlie is gone, but his lessons live on: it is possible to log in a sensitive, thoughtful way; and a man is perhaps his most manly when he is gentle and kind, when he is observant, listens, and cares. As for the forests, these eastern forests require the thinning once accomplished by regular fires. Adopting the traditional ecological maintenance practices of the First People, the BMFP would use light controlled burns. But for the present, much thinning must also be done by selective logging. And for logging to pay, there must be mills.

Boyd Britton credits Susan Jane's persistence, intelligence, and commitment as the force that held the Blue Mountains Forest Partners together throughout the collaborative's growing pains, as she calmly but clearly explained ecological realities while striving to hear and understand the needs of concerned citizens. I asked him why he felt he could trust Susan Jane when he first approached her to get the ball rolling that fateful day at PSU. Commissioner Britton said, "It wasn't a matter of trust. I guess I just have no shame."

Bonnie, Boyd's wife, corrected, "It isn't that he has no shame. He just doesn't fear rejection."

"I'm in business, in politics, and married," the commissioner twinkled. "I *can't* worry about rejection." And then he told the story about how he had courted his wife, the prettiest girl in the high school, and won her because everyone else was afraid to ask her out—afraid she would say no.

Susan Jane's work for the Blue Mountains Forest Partners was officially recognized early in 2012 when she won the Bridge-Builder Award given by Sustainable Northwest and Rural Voices for Conservation Coalition, only the second time in eleven years the award had gone to an environmentalist. Soon thereafter, Secretary of Agriculture Tom Vilsack chose her from among two hundred candidates to be on the twenty-member Federal Advisory Committee for National Forest Management, and reappointed her in August of 2014. Like the Blue Mountains Forest Partners, the committee is made up of a diverse group representing outdoor recreation, elected officials, timber companies, and conservationists. The group works at the national forest level toward collaboration that encourages rigorous research as the basis for decisions and supports rules balancing the needs of the ecosystem with those of human activities. The Trump administration chose not to renew the committee's charter, but before disbanding, it published a sixty-nine page document of final recommendations, a Citizens' Guide to clarify the planning process and let all citizens have input on how their national forests are managed, and a guide for state, local, and tribal governments. All stress the importance of shared stewardship, scientific basis for decision-making, and the importance of public participation. The citizens' guide helps demystify processes and opportunities, encouraging public involvement. "When concerned citizens work together," Susan Jane says, "we are at our best and strongest."

Western Environmental Law Center, where Susan Jane is staff attorney and Public Lands Coordinator, focuses on enforcing environmental laws, but it fully supports Susan Jane's work to find collaborative solutions and has recently broadened that work, adding attorney Pam Hardy, a collaborative negotiator, to the team. As WELC likes to say, "We use every tool in the box to protect and restore the wildlands and wildlife of the American West."

No longer at Malheur Mill, Mike Billman now represents Grayback Forestry, consulting and working with collaboratives in Malheur, Wallowa-Whitman, and Umatilla National Forests. The hope is to increase the scale of restoration, gaining a parallel increase in jobs. Born-collaborator and Mike's friend, Tim Lillebo of Oregon Wild was felled by a heart attack while shoveling snow in February of 2014. In Mike's new collaborative work, he will not be replacing Tim. No one could, Mike says. But Tim's spirit will be with Mike to help save the yellow-bellies; to protect watersheds for clean drinking water, for irrigation, for fish and wildlife; to restore the ecosystems for the hunter, the fisher, the hikers, and those who meditate or enjoy their adult beverage by the river.

Nearby Warm Springs, Paiute, and Umatilla tribes have not been actively involved in the Blue Mountains Forest Partnership, but likely are pleased to see the forest restoration and management working toward the conditions and methods of pre-European contact days. All three tribes participate in forest work through the Forest Service, with whom they have a good relationship.

The increase in scale and tempo of restoration appears to be what is needed now. A challenge common to environmentalist and lumberjack alike is the changing climate. Hot, dry weather produces water-stressed trees that become the beetle's playground and fire's fuel. As the climate warms and dry years become the norm, fires—not natural brush-killing fires, but devastating, big-tree-killing, home-threatening fires—will be more common. If the pace of restoration increases sufficiently, removing smaller more fire-prone trees and giving more room for fire-tolerant

Tim Lillebo 1952–2014. Forever missed.
Photo courtesy of Oregon Wild.

ponderosas, the greater resilience of the forest could provide refuge for various species as they move to more northerly climes in search of hospitable habitat. The work of the collaboratives becomes ever more urgent.

The benefit of collaboration for the economy and ecology seems self-evident. But not to be overlooked is the benefit to the collaborators. Lives have been changed. Hearts have opened. Friendships blossomed. From beginning to end, individuals learned to trust and respect the Other—people who had different backgrounds, different expectations, different goals, different looks and outlooks.

Working together, listening to each other, folks from divergent walks of life recognize common needs. With new information, a person can change her mind or direction without compromising her principles. Understanding the perspectives of others gives new personal freedom as well. You aren't afraid of someone you respect. And in collaboration, things can be accomplished that could never be done alone or in conflict. As John Shelk says, "If you bring people of goodwill together, they're going to solve 70 percent of the issues out there."

The knot of a problem becomes ever tighter when pulled from opposing ends, but worked at from the center, many hands can loose the snarliest tangle. Or together they may unravel the threads and weave a new and vibrant fabric.

Community

2

On the Road

Spring 2015

I SHOULDER MY BACKPACK UP NARROW STEPS and find my seat on Amtrak's Coast Starlight, heading first to Los Angeles, then to Flagstaff, Arizona, where our daughter Erika and her family live. From Flagstaff, I'll borrow a car and drive to Colorado.

Western Environmental Law Center's Hillary Johnson had told me about a Colorado river valley, renowned for fine wines, organic produce, art, and outdoor pursuits, whose residents had joined forces to protect their health, homes, and livelihoods from the threat of oil and gas exploration surrounding the valley. I want to see the place, meet a cross section of the people involved, and understand how a scattering of folks in a remote valley on the west slope of the Rocky Mountains could contest a powerful industry and government agencies.

We leave Eugene at 5:00 p.m., so, with sunset before 8:00 in mid-April, we don't get out of Oregon before dark. My Colorado exploration is many miles and several days away. Now it's time to settle in, let the train rock me to sleep, and enjoy California's Central Valley in the morning.

It's always a bit of a start, going to sleep in western Oregon's dense, moist green and waking in open, dry, sunny California. We roll by artichoke fields, by cabbages and broccoli; by flood-irrigated rice paddies with birds bunching in pools and exploding into the air; by cattle lying in fouled pens or standing chuck to chuck to chuck; by orange groves with their glossy dark green leaves; by peaches and almonds, past bloom now, their young fruit not visible from the train window. Toward noon come the magnificent hills—the Transverse Range—seeming to me the site of every Western film I ever saw. We snake through the Grapevine, and eventually we're in Los Angeles. Here, the semi-tropical plants thrill me, tickling memories of my years growing their Lilliputian replicas in tiny pots in northern commercial greenhouses.

In LA, I board Amtrak's Southwest Chief, heading east toward Flagstaff. Beyond San Bernardino, an hour along the road, we begin crawling up the mountains. We climb 2,743 feet in just twenty-five miles, coming out at 3,811 feet and trading Los Angeles palms for Mojave's scattered Joshua trees in wide stretches of barren, wind-blown land. This is the high arid and semiarid country where I will be traveling for the next couple of weeks.

After another night of swaying sleep, now thirty-six hours and about 1,300 miles from Eugene, Amtrak deposits me in the ponderosa woods of Flagstaff, Arizona, elevation 7,000 feet. I breathe in the crisp frosty morning air and fall joyfully into the arms of my daughter.

I unwind a bit, then, loath to leave my family but antsy to get on the road, borrow their 2001 silver Honda CRV. Heading northeast on the "Mother Road," aka Route 66, named and made famous by John Steinbeck in *The Grapes of Wrath* (1939), I sing at the top of my lungs, "Flagstaff, Arizona. Don't forget Winona . . . Get your kicks on Route 66." But I'm not going the right direction to make the song work, and I forget most of the words anyway.

Driving solo in someone else's car, on unfamiliar roads in unfamiliar states, with no GPS in the car or in my pocket, I am hopped up. Excited for the adventure, buzzed from the risks of not-knowing: a 450-mile trip, no picture in my mind of the destination, wanting to locate my camping place before dark. But beyond that hyper state, I am eager to reach my ultimate objective and spend time in the North Fork Valley of Colorado's Gunnison River.

At about eight hours from Flagstaff, head buzzing, bottom and joints aching, I stop for my second time. I am skirting the top of a canyon and with my whining body as an excuse, decide to check it out. It's hard to describe the sense of awe and tinge of terror I feel looking deep into the jagged maw of Black Canyon. Nearly vertical stygian spires crowd the dark and narrow gorge where, perhaps 2,000 feet below, a silver ribbon flows as far as I can see inside its snaggy black womb. This shiny wriggle is the Gunnison River, an exciting discovery. That means I should be able to find the valley of its north fork before dark. The river descends steeply—I read later that it drops an average of 43 feet per mile over the length of its 48 miles, and 240 feet per mile in the area of its most extreme descent—letting me preview the descent I would soon be making into the valley. But I linger, chest thumping. Legend says that the First People of this area never camped and rarely even traveled in the Black Canyon out of fear, superstition, or respect. I understand.

I pull myself away and back to the car. I need to find my campsite, and I want to make a plan for tomorrow and the following days. Citizens of the Gunnison's North Fork Valley are quite diverse in both history and occupation. Some families have lived there for generations. Others discovered the area in the past few decades. The demographics have changed recently. Until just a few years ago, there were roughly equal thirds of coal miners, farmers, and retired folks. Now mines are closing; more young people are coming, seeking sustainable lifestyles, and people recovering from potentially devastating illnesses are choosing the valley for its healing properties of clean air, clean water, and clean food.

Stereotypes would have you wonder if old-timers would resent newcomers, if coal miners shun the artists, but what I have heard and read says that is not so. I wonder why. During the week, I'd like to talk to a diversity of residents. I also want to find out as much as I can about what makes the area special, and if possible, reconstruct the story of the oil and gas threat.

About fifteen minutes later, a sign says Crawford is just ahead. Greening grassland attests that I'm already about a thousand feet lower than at the canyon rim. Cattle graze or lie about, and a small group of horses romp, tossing their manes. A ranch house, corrals, a background of mountains, and a half-dozen storefronts tell me that this indeed is Crawford.

The three little towns in the North Fork Valley constitute what some call the "Golden Triangle." Crawford, population 410, which I'm now driving through, then an eleven-mile straight shot to Hotchkiss, population 944, above which I will camp. If I should turn hard right from Hotchkiss and travel nine miles, I would find Paonia, population 1,451, where I will go tomorrow.

Crawford, Hotchkiss, and Paonia nestle among organic farms, orchards, ranches, and vineyards in the North Fork Valley, surrounded by mesas and mountain ridges mostly in public ownership—wilderness areas, national parks, and Bureau of Land Management (BLM) land. Lakes and streams charged by snow in high mesas cut through near-vertical canyons, producing lush flora and fauna in surprising diversity. Rushing creeks, popular with both local and visiting anglers, are home to brown, rainbow, and brook trout.

Numerous raptors soar and nest in the valley, which provides critical breeding, migration, and over-wintering habitat. Among them are ferruginous and rough-legged hawks, both "species of concern"—birds seeming in need of conservation, but with insufficient information available to be officially listed as threatened or endangered. Side canyons echo with the music of songbirds. The mountains invite camping, hunting, fishing, hiking, mountain biking, wildlife spotting, and species counting, as the valley lures farmers, ranchers, artists, and tourists.

I roll into Hotchkiss, mentally time-machined back to towns of my childhood. No big box stores, no chains, no billboards or flashing lights. If the cars on the streets weren't late models, I'd think I was in the 1940s or '50s. A bookstore and a historical museum beckon. But with a quick look to the right toward Paonia, I drive on, and up the hill to search out Colwell Cedars Retreat, where I will be camping.

I drive by orchards and vineyards, then wind my way nearly to the end of what turns out to be Redlands Mesa. The area is stunning. Idyllic. Orchards, vineyards, farms, all being watched over by the West Elk mountain range. I try to imagine the view interrupted by oil wells, but my mind refuses.

I turn at the sign *Colwell Cedars Retreat* and drive by cacti and sparse clumps of grass; by twisted, silver-trunked ancient junipers (the cedars of the retreat's name); by steep grades and flat, by gray and green. If I'm on the right drive, I *must* find time to explore. After minutes of awe, mixed with concern that I may have taken a wrong turn, my destination appears: the guesthouse and the home of artist Katherine Colwell and retired forester and author Joe Colwell.

Eager to get set up before dark, I decide to explore tomorrow. Myriad trails beckon, as well as the mysteries of Katherine's studio. But now I lay out my sleeping bag near the lodge, looking out to snow-blanketed mountains. At the Colwells' invitation, I use the bathroom of the guesthouse, as well as the house electricity, which gives access not only to greater comfort, but also to illumination sufficient that I can brush up on the history of the valley.

Sam Wade and Enos Hotchkiss were among the first Euro-Americans to discover this hidden paradise. On their trek from the Midwest in the early 1880s, they marveled at the mild climate, the canyon sides coursing with water that could be made available for irrigation, rich

Location of
NORTH FORK GUNNISON RIVER
Colorado

| 0 | 50 | 100 mi |

The North Fork Valley, showing the Golden Triangle: Paonia, Hotchkiss, and Crawford.
Map by Imus Geographics.

bottomland soils, and the protection afforded by the mountains and
mesas. Healthy native fruiting plants such as hawthorn, chokecherry,
and buffaloberry gave Wade hope that the land would support domestic
fruits as well. In the spring following his return to Missouri, he brought
back 200 apple trees, 10 each pear and apricot trees, 20 sapling peaches,
200 cherry trees, 100 grape vines, 1,000 blackberries, 100 raspberries, 12
currant plants, and 50 gooseberries, planting them on his farm, which
would later become the town of Paonia. Though he lost a third of his
plants the first winter, the rich alluvial soil did even better than he expected with the ones that survived.

Having begun with those orchards, the valley now produces 77 percent of Colorado's apples and 71 percent of the state's peaches. Wade

also brought in a load of his favorite flower, the peony. The Post Office department thought there were too many vowels in peony's botanical tag, *Paeonia*, but it was those flowers that gave the town of Paonia its name, shortchanged only by one vowel.

At about the same time that Sam Wade was planting orchards and building irrigation ditches in what would become Paonia, and Enos Hotchkiss was putting in his own fruit trees about nine miles southwest in what would become his namesake town, geologists discovered coal about ten miles northeast of Paonia. Though begun modestly, shaft mines became an economic driver for the area. By 1902, a railroad ran from the mines through Paonia and Hotchkiss, hauling not only coal but also fruit, making a name for the valley, and providing means for the whole state to taste the prize-winning produce of those early orchards.

CHC, Vintners, and the AVA

I crawl into my sleeping bag, pulling the space blanket tight around it, and gaze at the dense canopy of stars. The air is crisp, but I am cozy. My weary body aches, but I am wired. My head buzzes with visions of the day's venture, the history of the valley's settlement, and my hopes for what I want to accomplish during the week, all the while blocking the specter of oil wells from my mind.

I awaken to an extended sunrise show, the sky behind the West Elk mountains bathed in waves of color. Driving down the mesa, past the railroad tracks skirting Hotchkiss, and along the tracks toward Paonia, I feel the history of the past century and a half. The tracks that have hauled coal and fruit run to my left on the nine miles from Hotchkiss to Paonia. To my right, mountain ranges watch over verdant farms, pastures, and orchards. I almost don't see the turn, but there's the sign: *Welcome to Paonia*. I am immediately charmed by its turn-of-the-century quiet energy.

Jim Ramey, Executive Director of Citizens for a Healthy Community (CHC), greets me in the lobby of The Hive, an office building in the midst of other late-Victorian commercial-style buildings on Grand Street. Citizens for a Healthy Community is an organization dedicated to "protecting air, water, and foodsheds from irreparable harm." One of the earliest organizations to respond when the community was shocked with the peril of fracking near the valley's farms, vineyards, and outdoor recreation areas, CHC is the go-to group for this quest. Ramey, who lives in Montrose, fifty miles from Paonia with his attorney wife, Lindsey, was hired in 2012 by the then-all-volunteer CHC as soon as possible after the valley learned of the oil and gas leasing threat. He brought a background in energy and degrees in political science from the University of Cincinnati and in public administration from Ohio State.

Ramey recounts that CHC was begun in 2009 by Robin Smith and Daniel Feldman and their wives, who came to the valley a mere two years before the December 2011 BLM oil and gas lease-sale announcement. So I catch up with Robin, who explains that he and his CPA wife, Cynthia Wutchiett, had worked hard and lived frugally in Columbus, Ohio, squirreling away whatever they could spare, driven by the vision of leaving the hurry-scurry world of credit cards, sirens, and screeching brakes for an early retirement to simple and sustainable lives. When they discovered the North Fork Valley, they knew they had found their paradise.

Daniel and Joann Feldman studied permaculture and sustainable agriculture while they searched the western Rockies for the right spot to realize their vision of responsible and self-sufficient farming. They too found the North Fork Valley. But even as the last bricks were being laid on the Smith and Feldman dream homes, they heard rumors of gas development at the headwaters of the Gunnison River. If toxic chemicals leaked into the Gunnison, they would make their way into this, the valley of the Gunnison's north fork.

Recognizing the need to have strong and cohesive public protection for the valley, Joann Feldman and others began organizing concerned locals into a nucleus that would become Citizens for a Healthy Community. Robin Smith made contact with the Western Environmental Law Center (WELC) in Taos, New Mexico, to be sure legal help would be available if needed. The development at the Gunnison headlands didn't transpire, but when the BLM announced its imminent plans to open lands surrounding the North Fork Valley to oil and gas leasing, CHC began to mobilize.

Jim Ramey provides me with names of a diversity of locals as well as important activists. During the course of the week, I want to understand the oil and gas threats and how the people are working together to fight them, but first I need to get better acquainted with the valley and its residents.

I hop in my daughter's car to check out some of the wineries, but I'm also curious about the coal mines, and head a few more miles northeast of Paonia to see two, now closed, of the original three. Conveyor lines and silos remain but the area is tidy. No ugly piles of debris or waste coal culm are apparent. I'm surprised, although I had been told that the mines were good neighbors. I also learned that this is a hard anthracite, low in sulfur and cleaner burning than much other coal. I certainly see nothing to complain about as I look at the silent structures. That helps me understand the comparative lack of friction between the populace and the mines. Also, the mines were there from the earliest years of settlement. It wasn't as if they were latecomers, upsetting an entrenched way of life. And the mines paid their way.

Clearly the closing of two mines, as more customers turned away from coal and long-term contracts were not renewed, will have jarred the economy. That probably explains support for oil and gas development among some in county administration. In 2010, 950 miners worked at the three locations. By 2017, the area was down to the one mine, with 220 workers. That hurts both the jobless and the local tax base.

King Coal one roadside sign shouts, as I head back toward Paonia. Another sign reads, *Coal Keeps the Lights On.*

Just outside of Paonia, I take a right and start climbing toward Stone Cottage Wine Cellars. Brent Helleckson, now active on the CHC board, chose the Paonia end of the North Fork Valley to establish his vineyard and winery after he left his Boulder career as an aerospace engineer in 1994. Helleckson and his wife, Karen, dug foot-wide and bigger stones from the soil as they prepared to plant their vineyards. They used those stones to build their vineyard's namesake cottage, a charming half-buried wine cellar, a tasting room, and, eventually, their own home.

Helleckson is warm and welcoming, in an understated "aw shucks" sort of way. His slightly tattered, loose basket-weave hat shades his eyes, allows the filtering through of cooling breezes, and looks as though it may have become a permanent part of him. As he shows me around, he acknowledges, when asked, that grubbing those stones from the soil and then using them as building blocks was indeed a bit of an effort. But he volunteers nothing about the hours and sweat and blisters. He seems a person to see possibilities rather than hardships.

The Hellecksons moved to the North Fork Valley to live close to the land and raise their family sustainably. Buildings, vineyard, harvesting, wine-making, and bottling, all are family projects. It's the lifestyle they were seeking. The North Fork Valley is where they found it, and where they'll do everything in their power to maintain it.

Stone Cottage Cellars and their neighbor Terror Creek Winery both sit high above the valley. At over six thousand feet elevation, the views of the valley, nearly a thousand feet below, and the snow-covered mountain ridges beyond make me gasp, step back from the edge, then forward to look once again. Sort of a one, two, three dance step. My urge to look (*Ooooh! What a view!*) bubbles as intensely as my weak-kneed acrophobic response.

Another local vintner, Eames Petersen, learned wine-making in Spain and brought the old methods back to the States. He planted a vineyard near Paonia in 1994 and by 2012 his Alfred Eames Cellar label was already outshining French wines at juried tastings.

The elevation plus the circulation of warm air running up the valley at night, and cool air rolling off the mountains by day, provide a unique climate for the production of fine wine. Local lists include Chardonnay, Riesling, Gewurztraminer, Pinot Gris and Noir, Petite Pearl, Gamay, Cayuga, and Chambourcin as wines ideally suited to these high and dry vineyards.

With eleven wineries making wine from some of the highest vineyards in the northern hemisphere, the North Fork Valley's West Elk (named after the nearby West Elk Mountains) is one of only two American Viticultural Areas (AVAs) in Colorado. To qualify as an AVA, vineyards must lie in a geographically defined region with distinguishable characteristics such as soil, elevation, and climate. Surrounded on three sides by mountain ranges and mesas and on the fourth by desert badlands, the North Fork Valley is clearly defined. Three hundred sunny days a year fill the fruits with sugar; cool nights capture acidity; rich but well-drained soils make for unique crisp and flavorful wines, gaining the West Elk AVA justifiable fame.

These vineyards provide an obvious argument against encroaching oil and gas exploration and development. The American Viticultural Area designation is not randomly awarded. North Fork Valley is undeniably a distinctive area. West Elk wines are becoming well known and sought after, and the wineries have become tourist destinations. Wine sales and tourist visits both help the county's tax base. Clearly an oil or gas well in sight of the tasting room would not be a welcome neighbor from an aesthetic point of view. Add to that the industry's water demands and the potential for pollution, and protecting the area from oil and gas development would seem like a no-brainer.

Organic Horticulture

Next I want to visit organic farms and orchards, preferably calling on both the old-timer and newcomer categories. Mark Waltermire's work and background are captivating. He came to the North Fork Valley with degrees in biology and environmental studies, plus years of agricultural and teaching experience in Pakistan, California, Montana, and Massachusetts. In Massachusetts, he ran an education and food-bank garden involving people of many ethnicities. Education had always been an important aspect of his life, but happily, the diverse cultural component also introduced him to a variety of flavors and crops unfamiliar to most American markets.

In 2005, with his wife, Katie Dean, and their two young sons, Waltermire set up his sixteen-acre Thistle Whistle Farm outside of Hotchkiss, where he is known for the outstanding flavor of his produce (especially the hot peppers), its wide variety—the more unusual the better, as long as it's luscious—and his educational programs. Thistle Whistle produces 200 varieties of heirloom and specialty vegetables and easily sells all to nearby restaurants and, especially, directly to consumers.

Mark invites me to sit with him at a tree-shaded picnic table in a meadow. He wears a brown open-collared work shirt and jeans. A brimmed cloth hat shades his earnest and engaged blue eyes. His trimmed brown beard is lightly frosted. I feel immediately at ease.

Mark's two tall, blond sons and one of his interns soon join us. Dedicated to empowering others to live well, Mark runs an apprenticeship program on his farm, training future sustainable farmers. The interns learn as they work, as well as in classes and by taking part in on-site trials, such as an ongoing one exploring mycorrhizal effects on fruit trees.

Mycorrhizae are fungi that connect with roots of trees and other plants, extending the effective root network, thus giving the plant far greater access to water and nutrients. Trees reciprocate by sharing with the mycorrhizae the sugars they have photosynthesized. The networking and natural relationships that Waltermire strives for in his farming seem replicated in his work with students, community members, and organizations such as CHC. Networking, helping each other, working together, makes for potent communities as well as thriving gardens.

Waltermire provides his interns lodging, plenty of fresh produce, information, experience, a small stipend, and the keys to a fulfilling future. He presents classes in organic gardening to the community and makes regular visits to local schools to talk with kids about growing their own food. The students eat tomatoes fresh from the garden in the fall and save the seeds, which they plant the next spring. If they take their germinated seedlings home and tend them well, the plants grow and fruit, giving students not only a thrill and a good lunch, but also an indelible lesson in a tomato's entire life cycle.

Students from colleges, as well as from local schools, make field trips to Thistle Whistle Farm, and sometimes are inspired to become future interns. English-language learners from Delta, twenty miles away, come to spend the day discovering what plants grow on the west slope of the Rockies, how to grow them, and healthful ways to prepare them, as they practice this new language. Adults and children new to the area—many from Mexico, but some from as far away as Myanmar—hear, discuss, and take their notes on gardening information in English, with help when needed. They go home more confident in gardening, nutrition, and language. At the same time, Mark says he benefits by learning about their favorite varieties of vegetables and herbs that are new to him. He then adds delicious new flavors to his garden and to his table.

Thistle Whistle dedicates one area of the farm as the Sauce Plot, a space for kids to grow produce and herbs for the Kids Pasta Project (KPP), yet another example of people helping people. Local students learn about organizing, advertising, preparing, and serving a pasta dinner, with as many ingredients as possible locally grown. Local people—averaging about forty per dinner—pay for their meals and enjoy an evening out, visiting with friends. The profits from the KPP dinner go to a local organization. The organization benefits; the kids learn more about their community as well as the skills of putting on and serving a public dinner. In the summer, Thistle Whistle Farm sponsors the Sauce Plot Kids Camp.

Though the KPP cooks for various events, home base is at Edesia Community Kitchen, "where local food meets creativity." Edesia is a farmer's market, event space, and cooperative kitchen, where farmers and other entrepreneurs can create value-added products from homegrown produce, even if they don't have the required commercial kitchen.

Thistle Whistle is one of many organic farms in the North Fork Valley. In fact, the North Fork boasts the largest concentration of organic farms in the state. The climate has a lot to do with the valley's allure: Lots of sun. Clean air and water. That good, natural air conditioning I learned about at the wineries, with warm valley air lifting at night and cool mountain air relieving daytime valley heat. Another likely appeal is the complete lack of competing development in the area. Local attitudes that value clean food would be a clear enticement, as farmers need appreciative customers. And, of course, the farmers support one another. People of like values and skills share camaraderie and information.

A major attribute of the area is the quality of its soil. Mancos shale underlies the North Fork Valley, giving the soil above essential water-holding capacity plus a unique flavoring quality. Mark Waltermire says that of all the places he has farmed, his Thistle Whistle produce tastes the very best. He seems to have found the ideal spot to grow his own deep roots as well. But a potential weak link is water. Waltermire has a good little canal, its pure water fed by snowmelt. But he says the supply could become iffy with capricious snow levels, more and more prevalent with the changing climate. And now there is the added worry of a polluting industry that demands five or more million gallons of water per well coming to town.

Leaving Thistle Whistle

I gratefully accept the sample of goat cheese Mark offers, admire his goats and wandering chickens, and say goodbye, wishing I could stay longer, see his zillion varieties of tomatoes and peppers, and watch classes and field trippers discovering the magic of gardening.

On the way back to my campsite, I pay a brief visit to Wink Davis of nearby Mesa Winds Farms. Also organic, Mesa Winds has six acres of grapes, both table and wine, fourteen acres of pear and apple orchard, asparagus, assorted other vegetables, and Baby Doll Southdown sheep. It's a beautiful farm, and he's a charming man. He mentions that he'll be busy on the weekend, attending as much of the Climate Colorado Conference as he can squeeze in. I hadn't heard of the Climate Colorado Conference, but make a note to check it out, then head up the hill to home base. Before conference time, I still want to find a multigenerational farm. I wonder how the "old-timers" will look at the oil and gas threats. They learned to get along with the coal mines after all. And I wonder if I will see old-timer versus newcomer dissension. But that's for tomorrow. Now I'm eager to investigate Colwell Cedars Retreat.

Colwell Cedars

Delighted to have time before dark to exercise and explore, I set out on one of the many trails Joe Colwell has developed (thirty-one named trails, I read later, for a total of five miles in five distinct topographical areas). Immediately, I am taken with the old junipers, their long-dead silvery branches intermingled with the alligator bark of their living parts. I walk through hip-high sage, fragrant in the sun, through tufts and clumps of Indian ricegrass and needlegrass. Here and there, thick dark green mounds of yucca push their swordlike leaves upward. Then I plunge down a trail into startling green, picking my way across and beside a fast-moving stream. Tender emerald watercress carpets the ground, and new spires of cattail thrust toward the trees.

Later I find out that all of the creeks and wet areas along trails with names like "Laughing Water," "Wetfeet," and "Ruby Springs" are fed by springs in the hillside. I could happily sit here and listen to the sounds of water and birds, but eager to move my body and see as much as possible, I scramble back up the hill. Boulders covered with patches of lichen in shades of white, gray, mustard yellow, and rusty orange lie about near sculptures that are juniper logs and branches. Then I run to admire claret cups—circles of large, glowing, deep red flowers on a low-growing cactus. I've never seen claret cups blooming in the wild before, and I get out my camera. I wander a bit, vowing to save time for more each day, and then tap on the Colwells' door.

Being a "plant person," occasional teacher, and current writer myself, I feel I know where in my mind to file Joe's Forest Service, trail-building, environmental education work and book-author background, but Katherine's work is a complete revelation to me. Though I had read that she was a fiber artist, I am utterly unprepared for her two-sided and three-dimensional embroidery. Her roomy studio is full of projects finished, begun, still being designed. I admire a sculpture-like piece inspired by poplar trees, a stretched-out-accordion embroidered book, pieces combining quilting, drawing, and poetry with embroidery. I am entirely ignorant of this sort of art and move from piece to piece in awe.

Among the diverse population in the North Fork Valley, in addition to farmers, ranchers, vintners, and orchardists, artists and other professionals who work remotely find their ideal nests. The valley now is rich in studios and workshops of traditional artists (weavers, blacksmiths, carvers, quilters), fine artists (potters, painters, photographers, and sculptors), writers, performing artists, tech artists, culinary artists, and creators of value-added products. Katherine Colwell seems to have found a home among kindred spirits.

In 2012, the North Fork Valley was named an "Emerging Creative District" by Colorado Creative Industries, a division of the Governor's Office of Economic Development and International Trade. In June of 2013, that designation was boosted to Certified Creative District status—one more walk of life seeming to me incompatible with nearby oil and gas drilling.

How wonderfully serendipitous this choice of lodging turns out to be. Stretched out safely on the ground under the stars, I can appreciate the night sky and the mountains as I would not have otherwise been able to do. Wandering on the many paths on the Colwells' forty acres gives a beautiful capsule view of the plant diversity in the area. The North Fork Valley's unique position, surrounded by mountains and a desert, hosting a river and edged by magnificent Black Canyon, this particular convergence of ecosystems, this soil, this "air conditioning," makes it a special place. Adding the relative inaccessibility of the area, perhaps the North Fork Valley provides a refugium: a place where species can find refuge even in stressful times. If so, what a precious gift as the climate provides ever more stress. And how extraordinarily essential to protect such a refuge from the destruction of development.

Ela Family Farms

I find my "old-timers" in Ela Family Farms and am able to catch Shirley Ela by telephone. Ela (born Shirley Philips) may have bonded with her family's orchard when, as an infant, she gazed at leaf patterns and shadows, listened to breezes, and smelled the freshly irrigated soil, all from her mini-playpen—a fruit box plopped beneath the trees—under the watchful eye of a hired hand. Ninety years later, the orchard is still her favorite place, this perfect spot of western Colorado, between the mountains and the desert.

Shirley is a third-generation orchardist. In 1906, as land began to open up and latter-day pioneers dreamed of new lives, her grandparents, Frank and Maggie Burns, packed their belongings, loaded their animals into a railroad car, and moved their family from Iowa to the west slope of Colorado's Rocky Mountains, near Grand Junction where Frank planted peaches and lined irrigation ditches with boards to direct water to his trees and prevent erosion.

Frank and Maggie's daughter Lois married Nelson Newton Philips in 1918. They bought land and planted pears, then apples and peaches. Shirley was their third child. Her childhood memories revolve around the orchardist's life: those beautiful trees, the horse team pulling the wagon with space left between boxes of fruit for a little girl to ride along, trekking across fourteen acres to deliver drinks to the packers by the time she was ten, joining her father on marketing trips.

The orchard's packing shed gave Shirley a job from the time she was a high-school freshman. Then to junior college, where wartime training prepared her for a stint at Boeing Engineering in Seattle, marking corrections in plans for the B-17 planes. From there, it was back to Boulder for a degree in sociology from the University of Colorado. Near her graduation, her high-school friend Bill Ela came back from the war, and "he looked pretty good in that uniform," Shirley remembers. They married and moved to Cambridge, Massachusetts, where he would get his law degree at Harvard.

But the couple agreed that the West was the place to raise a family, and Bill says, "I didn't just marry a wife. I married an orchard." They bought a farm adjoining the Phillipses' land, and Bill practiced law and became a judge in western Colorado. With her father's help and tutelage, Shirley ran the tractor, the irrigation, the packing shed, and the farm. By the late 1970s, development began to overrun their land near Grand Junction, surrounding it with subdivisions, roads, traffic, and noise, so in the 1980s, the Elas bought orchard property sixty miles southeast near Hotchkiss on Rogers Mesa, and Shirley continued farming. Topping off her long career, in 2011 she was honored with a Lifetime Achievement Award from the Western Slope Horticultural Society.

With an academic background in biology, environmental geology, soil science, and sustainable agriculture, Shirley and Bill's youngest son, Steve, became the fourth-generation farmer of the family. Shirley and Bill built a new house in the middle of the orchard where they will live out their lives. And that orchard is where Steve—active in organic research, on numerous horticultural and organic associations and commissions, newly elected to the National Organic Standards Board, and proud father of young track athletes who may become fifth-generation orchardists—plans to continue farming in the valley's good soil.

Today, the Ela Family Farm grows more than fifty-five varieties of organic tree fruits—cherries, peaches, pears, apples, plums, plus organic grapes and multiple varieties of organic heirloom tomatoes. Their website shares recipes and says that everything they sell or make—from fresh fruit to artisanal organic jams, jellies, fruit butters, sauces, dried fruits, and cider—they grow.

If there is any "old-timer/newcomer" strife, I didn't find it here. Thistle Whistle and Ela Family Farms collaborate. A weekly subscription of produce (Community Supported Agriculture, or CSA) from Thistle Whistle often includes fruit from Ela's. The valley boasts a vibrant organic community that knows sales directly to consumers are the best for their business, for the produce, and for the appreciative consumers. They communicate with each other, help each other, and together become stronger. Steve Ela, Mark Waltermire, Wink Davis, and Brent Helleckson are all active in CHC also, where with numerous organic growers and five hundred other CHC members, they fight to protect the air, water, and soil of their valley from all threats.

Valley farmers are already feeling the effects of climate change with earlier bud-break, warmer summers and falls, more erratic weather, more agricultural pests, but most importantly, changes in the watershed. The North Fork Valley gets very little rainfall, most of their domestic and irrigation water coming from snowmelt. Less snow is falling now and it is melting sooner. Some aspen groves are dying, with loss not only of their beauty and ecosystem communities, but also their shading of streams, exacerbating earlier snowmelt along with warming and evaporation of the stream's water. If irrigation ditches run out of water in August, farms and ranches could be devastated. As Colorado experiences increasing periods of drought, fracking (a technique blasting powerful streams of water, sand, and chemicals into the ground to release oil and gas) used twenty-six billion gallons of water between 2005 and 2013. Agriculture uses more water than does fracking, but oil and gas interests pay more than most farmers can afford and therefore have many eager sellers. "Water flows uphill to money," says an attorney from a Boulder environmental group. Even more worrisome than the amount of water used in this water-scarce land is that fracking wastewater—more than 27 trillion gallons injected back into Colorado ground in 2015, enough to fill 41,000 Olympic-sized swimming pools—has been found to contain numerous toxic chemicals, including arsenic and heavy metals.

Paradise Threatened

Next day, I head to Crawford to meet with horse rancher and Realtor Tom Stevens. We sit on the second floor of his lovely new home and look out expansive windows to a scene worthy of any coffee table picture book. Lush green meadows stretch to the rocky feet of mountain ridges. Beyond, a narrow opening in the rocks beckons for exploration. In the foreground, horses frolic and a small herd of deer seems to flow over the corral fence.

Stevens searched for the perfect place before he moved to the North Fork Valley. For twenty years a Realtor in Boulder, he was upset by the changes brought by extractive energy industries there. The moment he decided to move may have come as he advised a friend of the worth of her farm. She had come to him in desperation. Many people don't realize that they can own their home, their orchard, their ranch, but have legal rights only from the ground up. When the West was opened for expansion, the government often retained control of any minerals below ground level. Not having the mineral rights to her land, Stevens's friend was now surrounded by sixteen oil wells, one less than twenty feet from her house. "I can't stand it anymore," she said. "The noise, the smell, the dirt, trucks running back and forth . . . I can't use my land. I have no peace. I have to get out."

"And I had to tell her," Stevens said, "you won't be able to sell your land now. It is virtually worthless."

So Stevens went searching for an undamaged place where he could live an authentic life—one according to his personal values rather than one responding to outside forces—a life with trust in the future. His quest ended when he found the North Fork Valley. On the edge of the town of Crawford, he discovered lush pasture near an imposing boulder called Needle Rock. The perfect place for his horses, Stevens thought. And for himself.

He finally began to relax, shrugging off what had become an omnipresent shadow of impending doom, and dared, he said, to dream once again. So he signed the contract for this dream spread. But the ink was barely dry on Tom Stevens's commitment to the land when he and the rest of the valley awoke one blustery December day of 2011 to staggering news: The federal Bureau of Land Management was announcing the lease sale of 33,000 North Fork Valley acres for oil and gas drilling.

Mancos Shale

Historically, people had understood that the Mancos shale underlying the North Fork Valley—the very shale giving the overlying soil its water-holding capacity and unique flavoring attributes so appreciated by local farmers and their customers—was not conducive to drilling. Any oil or gas that might be in or under the rock would be next to impossible to reach using conventional extraction methods. But that was before hydraulic fracturing (fracking) came on the scene. In this process, and with the horizontal drilling and multistage system expected to be required for Mancos shale, two to eight or more million gallons of water, sand, and chemicals under pressures sometimes exceeding 9,000 pounds per square inch are injected down one to two miles below the surface and then horizontally a mile or more, fracturing rock to release its hydrocarbons. Sand particles hold the cracks open so that the natural gas or oil can be pumped up the well. Since 2005, 22,615 wells had been drilled in Colorado as of 2016, with nearly 2,000 in 2012 alone, each well directly impacting about nine acres of land, and with that,

The pristine North Fork Valley rejects fracking.
Photo courtesy of Smith Fork Ranch.

introducing concerns about water pollution, water scarcity, air pollution, dust and noise, roaring trucks and hundreds of miles of new roads.

Was the idyllic North Fork Valley, home to thriving organic farms, orchards, and vineyards and renowned for art, tourism, ranching, and outdoor recreation, to become just another industrial wasteland? Tom Stevens looked over his newly acquired meadows, focusing on his imagined future. He told me he remembered thinking, "Such a pristine place here. No way can I face watching my world destroyed again. I can't run. I've got to stop it."

Girding Their Loins

Wanting to understand how Stevens and the many other concerned citizens joined together, I head back to Paonia, in search of the folks who were conducting the show before Citizens for a Healthy Community's Jim Ramey arrived. I track down Sarah Sauter a few blocks away from The Hive, in a modest bungalow housing the Western Slope Conservation Center. This thirty-five-year-old nonprofit environmental group has a diverse mission, working with water issues, education, and recycling "to build an active and aware community to protect and enhance the lands, air, water, and wildlife of the Lower Gunnison Watershed." The BLM's announcement threatened an assault on many fronts, and seriously endangered watercourses and ecosystems. Clearly, it was time to involve the public.

Sarah Sauter had arrived in Paonia just a year before the ominous BLM headlines, as the new executive director of the Conservation Center. With a toss of her thick blond braid, she gives me some background.

The fledgling Citizens for a Healthy Community was buoyed by knowing they had the Western Environmental Law Center's legal team to call on, but needed the support of the community at least as much. So together, CHC and the Conservation Center put up a Facebook page called North Fork Fracking, with bright red lines surrounding potential oil leases on maps of the area. The BLM's proposed wells could damage the valley's drinking and irrigation water, jeopardize recreation in the mountains and purity of the streams, compromise habitat for elk and other wildlife, threaten countless livelihoods, and destroy the character of the valley. The two organizations sent out flyers, public service announcements, radio and newspaper releases, phone calls, and emails informing the public, encouraging them to pass the word about the threat and its possible ramifications.

Sauter tells me that as the Conservation Center and Citizens for a Healthy Community scrambled to contact, inform, and energize the community, a man with curly graying hair and a short-cropped beard appeared in her office. This was Pete Kolbenschlag, a member of Colorado's Environmental Coalition and owner of Paonia-based Mountain West Strategies, a company specializing in community-based approaches to energy, public lands, and environmental issues. Long active at the state level, Kolbenschlag had maintained a low profile locally, so, though Sarah welcomed his help, she didn't initially realize the resource and talent she was getting. It turned out that Pete was not only a top-notch organizer throughout and beyond Colorado, he also had deep experience with and knowledge of the workings of the BLM. I make an appointment with this extraordinary gentleman.

I find Pete back at The Hive, where I had first met with CHC Executive Director, Jim Ramey. He breezes into the room, seeming to be squeezing me into a very busy day. His eyes sparkle with stored energy and a hint of a private joke that belies his quiet but intense demeanor. He tells me that in spite of his BLM connections, the oil and gas lease-sale plans caught him, like everyone else, by complete surprise. When he first heard the news, he put a map on the wall with a bright blue pin marking his property, then plotted out the proposed leases. They were on all sides of that bright blue pin. He was surrounded.

Pete had been going through a hard time in his personal life, he says, and as he stared at the map, he could feel "the third shoe about to drop." Riveted to the image on the wall, anger welled up. *No!* he thought. *I'm not going to let this happen!* He took a few deep breaths, set his jaw, and went to join the Conservation Center and CHC. Together, they rallied just about the entire valley.

The Community Coalesces

I learn about that first overflowing community meeting after the BLM announcement through conversations with Kolbenschlag, Kyle Tisdel, CHC's environmental attorney from WELC's Taos office (plus a few websites, some video, and several newspaper articles), and now, as I sit in Sarah Sauter's office:

January 4, 2012, more than five hundred locals gather at the Paonia Junior High School gym. Maps taped to the walls show the locations of the proposed drilling parcels. Tension fills the room, occasional voices exploding over the background buzz.

Can you believe this? This one is right by the school!

Here's one by the reservoir!

Oh God! Two are practically in my backyard. And three more in my view of the mountains!

Farmers, doctors, ranchers, artists; all professions, all ages crowd onto the gym bleachers and pack the chairs set up on the basketball court. Dozens more shoulder into standing room in the back. Residents are anxious, confused, angry, determined.

Panelists give an overview of the situation. Tension builds; agitated voices rumble. A panelist mentions that it takes 1,000 to 1,500 truck trips—huge trucks—to develop each well, reminding people of the noise, dust, congestion, and road wear those trips would produce. Someone points out that each gas well requires up to eight million gallons of water. *We've been forced to scrimp on water already!* a voice in the crowd exclaims, as recent droughts have made farms, orchards, and ranches struggle to find enough to get by.

CHC chairman Daniel Feldman shares with the public what the organizing groups have found particularly infuriating: the resource management plan on which the BLM is basing its leasing parcels was adopted in 1989. "Over two decades ago! Most of the organic farms and wineries, as well as some of the schools weren't even here when it was written. The valley has changed since 1989, and the BLM needs to know that. At the very least, they should be working from a current plan."

Then Sarah Sauter's AmeriCorps VISTA volunteer presents the *tour de force* of the evening, a virtual flyover of the area, using Google Earth mapping. As neighbors in the jam-packed gymnasium "fly" over their valley, seeing the reality of wells in their watershed, their viewshed, their and their children's lives, it is as if five hundred people have gone mute, total silence broken only by an occasional gasp or sob. Lease parcels completely surround the valley.

Kyle Tisdel looks out at the crowd, gratified at their engagement and determined to do everything in his power to support them. It was for moments exactly like this that he had felt called to public interest service: using the power of the law to support the needs of the people. Money had never been his motivator. The people's number-one need is protection of the environment, and climate change, the environment's gravest threat, is inseparable from development and use of fossil fuels and energy.

Tisdel explains the legal framework for challenging the bureau if they won't respond to citizen complaints. He says it is WELC's goal to assure that critical areas are permanently protected and that no shale gas development happens without environmental safeguards to protect both human and wildlife communities. Someone in the crowd shouts, "Who is it that wants to drill? If there's digging going on in my backyard, I ought to know who's holding the shovel."

BLM policy is to keep the identity of the nominators (persons or interests proposing to develop the lease) confidential, Tisdel explains. "But we agree with you. You should know," and he goes on to tell the crowd that WELC will be filing a Freedom of Information Act (FOIA) request to find out exactly that. All federal agencies must comply with FOIA. There are exemptions to the type of information that can be disclosed, but those should not be a problem for this case.

Heads seem to be held higher throughout the gymnasium, as the promise of legal support steels a growing resolve to fight for what each person present, individually or as part of the community, loves. If any drilling advocates are present, they hold their tongues.

Kyle Tisdel, ready to assist.
Photo courtesy of Western Environmental Law Center.

For four hours, community members ask questions or contribute information and share their personal stories and concerns. Someone says, "I get that the BLM needs to provide for multiple uses, and mineral extraction is a use. But how can the Bureau justify permitting a new use that will destroy other *existing* uses?"

Pete Kolbenschlag, speaking of huge leasing parcels at the base of Mount Lamborn, the summit dominating the valley, points out that those parcels are on geologically unstable land and right under avalanche shoots. Someone adds that fracking causes seismic activity, and wonders what that might do to coal miners working in underground tunnels.

Steve Ela tells the group that Ela Family Farms has put over a million dollars into the farm and they've been organically certified since 1995. He reminds folks of the many organic farms, grass-fed beef ranches, and micro-cheeseries in the valley that depend on clear water and a clean environment. The organic label and the quality of Ela's apples and juicy peaches—like the produce from the several dozen other organic growers in the valley—would be in serious jeopardy with resultant air or water pollution.

Then Theo Colborn, now 84, takes the stand. To thunderous applause and a standing ovation, her ever-ready smile, steely-gray cropped hair, and wide eyes as bright and penetrating as they must have been at age 20, the revered local scientist looks out at her friends and neighbors. Her voice passionate, she lists harmful impacts on human health of shale-gas drilling's toxic by-products. These include brain, nervous system, and immune system effects; endocrine disruptors; carcinogens; chemicals affecting the G.I. tract, the cardiovascular system, kidneys, liver, and metabolic processes, plus several chemicals that are genotoxic. Nationally controversial but fearless, Dr. Colborn was among the first to bring an awareness of endocrine-disrupting effects of even very small doses of chemicals. As important as it had been for her to ferret out these effects, she now feels an even greater urgency to share the information.

Theodora Decker Colborn

Theo Colborn, born Theodora Emily Decker, March 28, 1927, in Plainfield, New Jersey, was honing her love for the outdoors at about the same time that Shirley Ela was bonding with her Colorado orchard from her nursery-box cradle. From earliest childhood, Colborn was fascinated with birds, the natural world, and river water as she tried unsuccessfully to find answers to "Why?" and "How?" By the time she hit high school, she revolted against prescriptions directing girls to courses that prepared them for secretarial or teaching careers, and opted instead for "boys only" classes in science. There, she excelled.

Theo was awarded a four-year scholarship to Rutgers College of Pharmacy, and after getting her bachelor of science degree, she married Harry Colborn, a WWII vet and fellow pharmacy student studying on the G.I. Bill. Together they had three pharmacies and four children, taking the family birding, camping, and exploring the out-of-doors as often as possible. Looking for quieter and wilder lives, in the early 1960s the Colborns sold their pharmacies and headed west beyond the ridge of the Rocky Mountains to land near Hotchkiss. But before long, Theo became uncomfortable selling pharmaceuticals, as she thought about the drugs' possible side effects. At about the same time, she noticed ill health in locals who drank water from the Gunnison River where it had been polluted by coal mining. She wanted good data to fight for the health of the people and the rivers, but she couldn't find any. So she took a job sampling waterways.

Early in her sixth decade, Theo went back to school, spending her summers doing field research at the Rocky Mountain Biological Laboratories in Crested Butte, studying the effects on aquatic life of pollutant residues from coal mining. In 1981, her master's thesis about using aquatic insects to measure the trace elements cadmium and molybdenum was accepted by Western State College of Colorado in Gunnison and, at age 54, she had a master's degree in freshwater ecology.

But she wanted to learn more. So she began studies in zoology at the University of Wisconsin, Madison, and at age 58, was awarded a PhD in zoology with distributed minors in epidemiology, toxicity, and water chemistry. After a stint at the White House in the Office of Technology Assessment, Colborn worked with the newly merged World Wildlife Fund and Conservation Foundation, doing research on contaminants in the Great Lakes.

Many scientists had worked on the Great Lakes, but they had not compared data. At a time when most scientists were reductionists, Theo was a synthesizer. She gathered the literature and produced a spreadsheet with animal names on the y-axis and observed effects on the x-axis. Resulting information was stunning. She found numerous species with reproductive and immune system problems and behavioral, hormonal, and metabolic changes even when the toxin dose was well below that considered safe.

Whole populations were declining or disappearing; many species showed birth defects or wasting before or after hatching; scientists observed ineffective mating behavior, thyroid problems, fish and amphibians with both male and female organs. The worst effects were in the young or in the unborn fetus and embryo. Often, an adult predator—avian, mammal, or other—eating fish from the Great Lakes would show no effects, but their progeny, if successfully born or hatched, would have any of a number of metabolic, neurological, or behavioral problems, and frequently would not live long enough to reproduce. One study reported that babies of human mothers eating Great Lakes fish two or three times per month had lower birth weight and had changes in their brains, visible on MRIs. In a paper explaining these effects, Colborn and others described how the toxic substances in the body block intercellular communication that would otherwise direct cellular migration and differentiation at early stages of development. These pollutants, even in very low doses, can affect bones, heart, nervous and reproductive systems.

The ultimate "knitter," as one colleague called her, "both of ideas and of people," and motivated by the Great Lakes experience, in 1990 Colborn convened twenty-one scientists from several disciplines to discuss environmental endocrine-disrupting chemicals and their effects. This work became part of the background for her 1996 book, *Our Stolen Future,* co-authored with Diane Dumanoski and Pete Myers. Written for the popular audience, *Our Stolen Future* has been compared to *Silent Spring,* both from its effectiveness alerting the public to a problem and for the virulent and abusive reaction its author received from Industry.

In 2003, at age 76, Theo Colborn returned to Paonia and founded The Endocrine Disruption Exchange (TEDX), an international non-profit organization for the purpose of "compiling and disseminating scientific evidence on health and environmental problems caused by low-dose exposure to chemicals that interfere with development and function." With a big-picture dedication to interdisciplinary work and passion for supporting women in science, TEDX is composed of women with advanced degrees in cognitive science, entomology, and integrative physiology. As well as looking at agricultural and industrial chemicals that have seeped or been dumped into waterways, Colborn and TEDX study environmental health effects of natural gas extraction, particularly those of fracking. The public isn't privy to the cocktails of chemicals injected during drilling, but of those that are known, many are toxic. This polluted water can come back to the surface or migrate underground to waterways or water-well sites.

A number of highly toxic gases in addition to methane are present at the wellhead, among them the "aromatics," collected in condensate tanks and delivered to chemical and product-manufacturing industries. These are the feedstock for much of what makes our lives easier and more hazardous: the plastics and pesticides, cleaning products and fragrances, fire retardants, pharmaceuticals, toys and electronics whose off-gassing can enter the womb through the mother's placenta and cause endocrine-driven disorders like ADHD, autism, diabetes, obesity, endometriosis, and early testicular cancer.

Theo Colborn says that the rest of her life is "dedicated to spreading the word that fossil fuels are not only linked to climate change, but also to the plethora of epidemics resulting from exposure to their end-use products."

Dr. Colborn's adult life epitomizes the effectiveness of working together. On the Great Lakes study, she brought scientists together virtually, sharing their results on her grid, even though they had not shared them in person. Then she gathered them face-to-face at the Wingspread

Theo Colborn, scientist extraordinaire. 1927–2014.
Photo courtesy of The Endocrine Disruption Exchange.

Conference Center in Racine. Next, she assembled bright women scientists from varied disciplines to create maximum diversity and intellectual rigor for her new research center, The Endocrine Disruption Exchange. Finally, through her book, papers, talks, and activism, she still acts as a powerful catalyst, educating and assembling the public to protect their own and the planet's health.

After Dr. Colborn speaks at the public meeting in the school gymnasium, CHC chairman Daniel Feldman tells the crowd that because of the flood of calls and emails the BLM has received, the comment period has been extended to February 9. "Go to the wall maps," he urges the audience. "Find every spot you care about—your homes, work, hiking or fishing places—and note every school, domestic well, irrigation ditch, or spring that could be damaged by drilling. Then write a letter to the BLM, being as specific as possible."

Though most of the people at the meeting are understandably focused on effects on their own businesses and ways of life, Dr. Colborn has given them a wider perspective—something more to think about as they write their letters to the Bureau of Land Management.

Action Accelerates

Fired up after the public meeting, the community immediately got busy. By February 9, the end of the extended comment period, they had written and sent more than 3,000 letters, plus made dozens of official protests and phone calls to the BLM. The law center filed technical comments on behalf of Citizens for a Healthy Community outlining legal deficiencies in the BLM's NEPA analysis for the lease sale (NEPA is the National Environmental Policy Act, a law from 1970 that requires federal agencies to assess environmental effects of their proposed actions before making decisions and requires the agencies to inform the public and give opportunity for public input) as well as their need to consider human health and climate risks of fracking.

Though presumably generating a revised version, the BLM was still working from an outdated Resource Management Plan that didn't deal with more than twenty years of change to the valley, so local citizens from business, agriculture, resource, and conservation communities joined forces to create their own alternative plan—one that would truly respect multiple uses, as the BLM claimed to do, including those both of the environment and of the community as it now existed.

Meanwhile, people got back to their lives. Orchardists and viticulturists had pruning to do; farmers prepared the soil for planting and checked their irrigation canals and ditches; the town of Hotchkiss readied for the annual sheepdog trials; ranchers gathered cattle for the spring drive through Crawford to lush higher pastures.

Citizens for a Healthy Community, until then a strictly volunteer organization, increased the size of its board of directors and advertised for an executive director. By spring, CHC had hired Jim Ramey as its first executive director. Ramey jumped feet first into the middle of the fray, just in time to learn that the Bureau of Land Management had denied WELC's Freedom of Information Act request for the names of the oil and gas lease nominators.

In mid-April 2012, as new leaf buds on vines began to break, Pete Kolbenschlag and a delegation from CHC headed to Washington, DC, to talk to Colorado legislators and to the US Director of the BLM. Legislators and agency heads in Washington should understand what was happening locally—the potential risks to the health and well-being of the citizens, the wildlife, and the economy.

Spring 2015

Now it's almost exactly three years later. I sit in Sarah Sauter's office spellbound by the story. *So then what?* I ask. She laughs. It was actually quite a surprise. On a sunny day in early May, as Sauter was striding down the street in Paonia, a woman she doesn't remember having ever met rushed up to her, threw her arms around Sarah, and sobbed, "We won! We won! Can you believe it? Thank you! Thank you! Thank you! I am so happy!"

It was news to Sauter, but it was true, to a point. The BLM had deferred the leasing of all twenty-two parcels. Was it the trip to DC? The legal presence? The public outcry? The community was giddy. *We did it! We did it!* You can *too* fight city hall!—or in this case, a federal bureau. Some began to relax. Some were hard at work on developing the community alternative for the existing and outdated resource management-plan revision. Some, though pleased for any delay, were all too aware that a deferral was not a cancellation. The area was still exploitable for the potential riches hiding deep within the ground. And in June, WELC filed a federal case to force disclosure of nominators' identities.

Climate Colorado

My visit with Pete Kolbenschlag takes much longer than he probably can spare. He glances at his watch and grins. "You should check out the Climate Colorado Conference while you're here. We can't just be against things. Come see what we are *for*." This is the same conference Wink Davis had mentioned when I was visiting Mesa Winds Farm, and it turns out that Pete is in charge of it.

Though the majority of folks in the valley strongly agree that no drilling should be allowed here, representatives of some local governments miss the budget-balancing income of the coal mines, and some residents muse that because they do appreciate having electricity, drilling may need to take place somewhere, and they don't want to be seen as NIMBYs—those folks who want important projects to be somewhere else, just "Not In My Backyard." But on Colorado's sunny west slope of the Rockies, there are nonpolluting answers to the need for energy, and a vibrant renewable energy program would bring jobs and local income. In early May, Pete's Mountain West Strategies presents a three-day Climate Challenge and Solar Fair hosted by Climate Colorado and Solar Energy International (SEI) to speak to some of those concerns.

I scoot over to the SEI campus Friday evening. After visiting information displays and the refreshment booth in nearby tents, I take a seat on the lawn to hear histories and missions. Founded in 1991, Solar Energy International is a nonprofit educational organization that provides technical training in renewable energy and sustainable practices to empower people, communities, and businesses worldwide. They give hands-on workshops and online courses in solar photovoltaics, microhydro, and solar hot water technology, working throughout the Americas as well as in Africa, Micronesia, and the Caribbean. At home later, I would discover that one of my favorite local organizations, Aprovecho, a nonprofit center that researches, demonstrates, and teaches sustainable living techniques, had staff who had trained with SEI and were in regular contact for information and support.

Climate Colorado challenges communities to find ways to reduce individual and community carbon footprints to net zero, and reduce water consumption by half. "We can't wait for the political system to respond," Pete says, "especially when Colorado loses. We are positioned to be global leaders." Climate Colorado's vision is that global problems like climate change can be tackled if they are approached as local problems with local folks working for solutions together. Many groups throughout the state are doing good work, Climate Colorado spokesman Robert Castellino says. Climate Colorado would like to help them connect and support one another.

The evening is mostly get-acquainted time. Meet and greet. Pick up information. Listen to music; nosh and drink a bit. Living off-grid myself, but far from being expert, I help myself to an armload of off-grid literature. I leave in time to get to home base before dark and check over Saturday's program on water issues and collaborating for local power generation. Not being pre-registered, I can't attend any of these, but I appreciate that they're happening and learn what I can about them. Sunday I stop by the Paradise Theater to take in the conference public matinee of the documentary *Merchants of Doubt* (2014). This is a chilling film about manipulation of scientific data to sell unsafe products to the public or to sow misinformation and confusion about subjects such as tobacco, sugar, or climate change.

In response to the closing of coal mines, its loss of jobs, and the hit to the economy, SEI began "Solarize North Fork Valley," a project educating and giving job training, as it partnered with local businesses, the school district, and the local rural electrical co-op toward the eventual goal of installing 120 megawatts of local renewable energy. SEI envisions a world powered by renewable energy and looks to the North Fork Valley as a model of that vision.

As businesses and homeowners in Paonia, Hotchkiss, and Crawford take part in the Solarize project or accept the climate challenge to work toward net zero energy and reduce water use by half, increasing numbers of the community understand that fighting fracking doesn't make them NIMBYs, nor does it mean adapting to a life without an energy source or jobs. Abundant jobs are available in the alternative energy sector. It's not jobs or the environment. It's jobs *and* the environment. As SEI continues its work, its reach extends throughout the nation and the world, teaching people technical know-how, spreading possibilities and hope.

Roller Coaster

In conversations with Jim Ramey, Pete Kolbenschlag, and Kyle Tisdel, they all brought up Thompson Divide, which rang no bells for me. So it was time to search out old newspapers and other reports.

In November 2012, Valley citizens learned that the White River National Forest leadership was drafting a fifteen- to twenty-year oil and gas plan. If residents could do their research and organize quickly, they would have a chance to weigh in.

Immediately north and east of the North Fork Valley, the 2.3-million-acre White River National Forest includes eight wilderness areas, with ten mountain peaks over 14,000 feet high. One edge of the forest climbs to the top of Mount Aspen. Throughout, it hosts cattle ranching, tourism, hiking, camping, hunting, fishing, and skiing. Called by Governor John Hickenlooper the "crown jewel" of the area, the 220,000-acre Thompson Divide is located in the White River National Forest. It runs through parts of five counties and holds 81 mineral leases, covering nearly half of its total acreage. Just west of the divide, oil and gas activity has turned tens of thousands of leased acres in Gunnison County alone from wild lands to potential or active industrial desert.

But before the extraction blight spread farther, the public had an opportunity to give input to the planning process. People were urged to tell the Forest Service, "No drilling in the Thompson Divide!" Locals were aghast and visitors incredulous that drilling could be considered in this rugged and flourishing landscape. Perhaps it wouldn't have come as such a surprise, though, if the populace had been aware of what had been going on earlier in Washington, DC.

When George W. Bush left Texas and his oil interests for the White House in 2001, energy exploration was high on his agenda. With the encouragement of his administration, the BLM, which manages all federal minerals including those in national forests, put millions of acres up for auction, often slighting both environmental analysis and public notification. The fervor of the moment encouraged what Wilderness Workshop attorney Peter Hart called a "lease before you look" mentality.

During this period of furious speculation, the only way to find out about a proposed lease sale might be to stumble on a note tacked to a bulletin board. The plan for leasing in the Thompson Divide might not have been discovered before drilling was active but for the sharp eyes of an attorney in the public interest environmental law firm Earthjustice. He saw a notice of an upcoming auction and spotted proposed leases in a roadless area. Joining with Wilderness Workshop, he filed a challenge, alerting others to the potential threat.

As environmental watchdog groups prevailed in challenges to the Department of Interior's Board of Land Appeals, the Bush administration worked on a new energy policy. When rules didn't favor the industry, they needed to be tweaked. Overseen by Vice President Dick Cheney, former head of fracking-fluid-providing Halliburton

Industries, and with major input from representatives of the oil and gas industry, the stated purpose of the 2005 Energy Act was to "ensure jobs for our future with secure, affordable, and reliable energy." It changed US energy policy by providing tax incentives and loan guarantees for energy production and exempting the industry from many public health and environmental laws. Thanks to the act, oil and gas producers could drill without being held back by the Clean Air or Clean Water Acts or the Safe Drinking Water Act. The Energy Act also created a loophole exempting producers from disclosing the chemicals used in fracking, and it directed the Secretary of the Interior to analyze a program for leasing out shale and tar sands on public lands, particularly in Colorado, Utah, and Wyoming.

And so the rush began, virtually unfettered, to explore, exploit, and profit from the seductive geology of the Thompson Divide, among other places. Residents from Aspen to Paonia, passionate to save the stunning divide from transformation to oil fields, plunged into the fight.

Focused on soliciting and sending letters to protect the Thompson Divide, CHC and residents of the North Fork Valley were blindsided when, in the middle of that same November, the BLM announced it was putting twenty parcels, about 20,000 acres of land in and around the valley, back on the auction block, scheduled for sale February 2013. Once again the timing of the announcement, and therefore of the requisite thirty-day protest deadline, coincided with the holiday season. And the Western Environmental Law Center was in the midst of preparing a federal court case to uphold the FOIA in response to the bureau's denying the request to reveal the nominators' identities.

As is required for any proposed agency action on public lands, the Uncompahgre Field Office of the BLM published the federally mandated environmental assessment of their proposed lease sale. An agency must assess what effects its project might have on the human environment and suggest alternatives if it sees the potential for a negative impact. If it recognizes no problems, it issues a FONSI (finding of no significant impact) and prepares to proceed with the project.

After North Fork residents, organizations, and governmental agencies sent more than 3,000 carefully researched letters enumerating the myriad ways that oil and gas development was incompatible with their lives and work, as well as with the health of wildlife and preservation of natural attributes, residents of the valley were incredulous when the BLM's Uncompahgre Field Office issued a FONSI.

"No significant impact" in the face of possible pollution of air, water, and soil endangering human health as well as the certification of organic farms, orchards, and vineyards and the quality of their products? "No significant impact" though drilling would disrupt breeding and migrating grounds of wildlife, clearly impacting fishing and hunting in addition to the wildlife itself? "No significant impact" in the potential destruction of the tourist business and property values? Not, of course, to mention "no significant impact" to climate change.

Though the document carried the name "environmental assessment," it seemed to completely neglect assessing any possibility of environmental effects of oil and gas exploration or drilling. With perhaps unintended irony, the bureau claimed that no analysis was required at that point, as there could be no environmental effects felt before leasing. It went on to say that if any should occur once work began, they could be mitigated, which rationalized the FONSI. "Nothing bad will happen," it seemed to say, "but if it should, we'll fix it." Therefore, whatever bad might happen would have "no significant impact." Clearly, the bureau was considering the leasing process to be the federal action, while the concern of the community was the effects of the fracking to follow, which is difficult to impossible to challenge once the parcels are let.

Thoroughly frustrated, the community sought the help once again of laws made to protect both the people and the environment that sustains them. Their attorney, Kyle Tisdel, sent a thirty-six-page letter protesting the bureau's assessment to Helen Hankins, Colorado's State Director of the Bureau of Land Management. The letter objected to the BLM working from an out-of-date resource management plan and therefore not considering the many changes to the area over the last nearly quarter century. It noted that though a number of nearby lease sales were proposed, no attempt had been made to assess their cumulative effects and stressed that *potential* problems must be identified and avoided. They must not be ignored, with the hope of correcting them after the fact.

With scheduled lease sales just two months away, the people hoped for a quick response to their protest, but by late January 2013, with the proposed date for the sales just a couple weeks away, there was still no word from the bureau. So it began to look like hardball time.

On January 25, as CHC prepared for litigation, CHC Executive Director Jim Ramey wrote to state and federal officials. He thanked Acting

BLM Director Mike Pool and Deputy Director Neil Kornze for meeting with North Fork's coalition of residents on their recent trip to the nation's capital and reminded them of the upcoming lease sale and the public's near unanimous concerns. He expressed the citizens' hope that they could continue to work with the bureau during the management-plan revision. Ramey concluded that CHC hoped litigation could be avoided and that the bureau would work with North Fork Valley residents to prepare a management plan reflecting the present reality of the community. "However," he wrote, "if litigation is the only option that we are afforded, our attorneys at the Western Environmental Law Center are prepared to file." Copies of the letter went also to US senators and representatives, a state senator and representative, the Delta County commissioners, and the Paonia town council.

And then miraculously, or so it seemed to some of the onlookers, February 6, 2013, one week before the proposed sale, the bureau deferred action on all parcels. State Director Helen Hankins said, "We've listened to concerns raised in numerous comments and public meetings. We are responding by deferring the North Fork Valley parcels at this time." Kyle Tisdel called this an example of how organizing and legal strategy can come together to reach a successful resolution.

The BLM finally acknowledged that their Resource Management Plan must be updated before they could rationally propose selling drilling leases. Jim Ramey expressed his group's delight that the bureau would be "slowing down and taking a closer look at the severe impacts of drilling." Not surprisingly, industry groups didn't share that delight.

Kathleen Sgamma of the Western Energy Alliance argued that farming and gas drilling coexisted in many places. West Slope Colorado Oil and Gas Association E. D. David Ludlam had been annoyed for some time that Ken Salazar, President Obama's Interior Secretary, required proposed drilling parcels to be identified early, the BLM to conduct reviews, and that public input be sought. That just adds work and slows things down from the industry point of view. He said that now the bureau's decision to defer takes an already bad situation and makes it worse.

But much of the North Fork Valley was overjoyed. "This is what we wanted," said Mark Waltermire, "a chance to get the management plan to reflect the valley's promise."

Then on February 13, more good news: US District Court Senior Judge Richard P. Matsch declared unlawful the BLM policy of concealing from the public the identities of those nominating public lands for resource extraction. Judge Matsch pointed out that knowing names makes bids competitive and competition elicits a fair price for publicly owned minerals. Not incidentally, knowing the identity of submitters can also help concerned citizens evaluate the submitter's history of environmental stewardship, giving the people a better basis to evaluate potential effects.

The judge's decision was a big deal. For the first time since the oil and gas boom began, the public's right to know was affirmed. "A precedent has been set," Jim Ramey exulted. "This is a victory for everyone who believes the government should do its business in the open." Though the FOIA case had specifically requested identification of the nominators in and around the North Fork Valley, the decision will affect BLM-managed public lands across the nation.

"Every community has a right to know what corporations are seeking to drill on public lands near their homes or where they recreate," said Kyle Tisdel.

While the valley cheered that decision, the committee of organic growers, vintners, ranchers, businesspeople, and conservationists put the finishing touches on their North Fork Alternative Plan and gave it its public debut. Chaired by CHC's Jim Ramey and Western Slope Conservation's Sarah Sauter, the "community alternative" would adapt existing BLM management tools to protect the North Fork's unique attributes, including prime wildlife habitat; delicate soils; domestic and irrigation water supplies; hunting, fishing, backpacking, and other recreational areas; as well as create tools to safeguard existing and emerging economies. The guiding philosophy was that even if oil and gas drilling must happen, it doesn't have to happen everywhere. Some places should be off the table. The multiple-use doctrine doesn't require destroying existing uses to favor the proposed use. And some things—water, air, children—are inviolate.

The BLM had until Monday, April 15, sixty days from Judge Matsch's decision, to appeal the order to make public the lease nominator identities. As the deadline drew near, there was still no word. But by mid-month, with the valley's orchards in astonishing bloom and farmers praying for no late freezes, BLM officials released what had, until then, been considered secret information. The nominators included Gunnison Energy, owned by Bill Koch and long associated with coal mining north of Paonia; Contex Energy Company of Denver; and by far the largest number of parcels, Denver's Baseline Minerals, Inc.

Tisdel and Ramey lauded both this precedent-setting decision and BLM'S cooperation in revealing names. The only problem was that Contex and Baseline are industry landmen, companies that research, negotiate, and bid on behalf of someone else, leaving still cloaked the identity of the company who would do the actual drilling. But the bureau seemed reluctant to revise their policy to reflect the court order requiring real transparency. In May, twenty-nine organizations would, with WELC's help, petition the BLM to update their policy.

Citizen Science

That same May, CHC, with input from the scientists at Theo Colborn's The Endocrine Disruption Exchange (TEDX), launched an innovative citizen-science air-quality sampling project. For a twenty-four-hour period twice a month, four times a year, volunteers would wear back-packs equipped with air-sampling devices as they went about their daily routines.

This is ingenious research, providing unarguable data. When the in-dustry gets complaints that oil and gas drilling pollutes the air and affects people's health, industry spokesmen might argue that no one can prove the pollutants weren't present before the drilling began, nor can they show that people actually breathe that particular air. This project speaks to both arguments. The backpack devices sample the air the volunteer is breathing, and through this project, a baseline of air quality can be established. CHC also developed a handbook on their methods to help other communities that wish to test for air quality. North Fork's first sampling was scheduled for the following September.

The year 2013 continued in a sunny upbeat vein: In June, the For-est Service withdrew drilling approval in the Gunnison headwaters. July saw the 67th annual celebration of Paonia's Cherry Days. In September, volunteers began donning backpacks, and air sampling commenced. Then the valley celebrated its bounty in the Harvest Festival, with mu-sic, poetry, classes in sustainable living, sampling of the valley's good food and wine, and dancing in the streets.

In October, the BLM agreed to consider the carefully crafted North Fork Alternative Plan while revising the Resource Management Plan for the area. Celebrating the decision, Brent Helleckson, owner of Stone Cottage Cellars and representative of the West Elk Winery As-sociation, declared that the Alternative Plan offers "reasonable recom-mendations to make sure we keep the North Fork the vibrant, beautiful, creative, agricultural hub that it is . . . [T]he North Fork plan would go a long way to protect investments this community has already made."

Amber Kleinman, a Paonia Town Council trustee, said, "People don't move to Paonia to live in an industrial zone." The NFAP is a "sensible and prudent, resource-based approach to protecting what is important to the town and our residents—clean water, incredible scenery, and a rural lifestyle."

Realtor Bob Lario added, "The North Fork Plan is very protective of the valley's resources and character, and that in turn is good for attracting investment and new business and protecting property values."

"We live here because we love the land and the place," Thistle Whistle owner and Valley Organic Growers Association President Mark Waltermire observed. "We need the clean water; we breathe the clean air. We want to raise our families in a farm field, not an oil field . . . [W]e are grateful and hopeful about this chance to make our voices heard. We love this place and will stand up for it."

More good news came toward the end of the year. Though the BLM administers subsurface mineral leases in public lands managed by other agencies as well as their own, they do so only after the managing agency authorizes the specific area to be leased. As they had earlier at the Gunnison headwaters, the Forest Service now closed large swaths of the Thompson Divide to further leasing and added protections to roadless areas. While savoring the moment, the divide's Wilderness Workshop and the Thompson Divide Coalition still hoped for legislative action to withdraw the entire divide from future leasing and to buy out or negotiate trades for existing leases.

Such progress could not come too soon. On November 14, 2014, Theo Colborn grabbed the attention of the valley populace with an urgent essay. She stressed the ubiquity of fossil gases, including the many aromatics, semi-gaseous liquids that escape from wellheads. She correlated increased human exposure through the past fifty years to a similar increase in hormone-driven cancers, infertility, neurological disorders, and a host of other effects. Even in minute quantities, these hormone-mimicking gases act on cell development from egg to senescence, affecting the structure, function, intelligence, and behavior of the individual. She pointed out that aromatics near wellheads have been found at more than three times the concentration known to cause birth defects. She reminded her readers that scientists know humans are causing climate change through the use of fossil fuels, but ended her essay saying that endocrine disruption is a far more imminent threat. Governments must act quickly, she said, "or too few healthy, intelligent people will remain to preserve a humanitarian society and create some semblance of world peace."

A month later, at age 87, Dr. Colborn died from lung damage that she had attributed to cadmium exposure received during her sampling of Colorado streams thirty years earlier. Though she is and will be sorely missed, her research, her writings, her courage, and her resolve will continue to inspire. Under the guidance of Dr. Carol Kwiatkowski, The Endocrine Disruption Exchange continues her work. I missed Theo Colborn by just a few months when I visited the North Fork Valley. When I was talking with Pete Kolbenschlag, he said he wished I could have known Theo.

I wish so, too.

April 2015, SG Interests and Ursa, two Texas-based gas companies with more than 40,000 acres of leases in the Thompson Divide, proposed trading that land for leases in Garfield and Delta Counties. Some saw a relief for the divide, but Peter Hart, Wilderness Workshop attorney, noted that the proposal included no protection for the area of vacated leases, which were illegal in the first place. *Why should the lease owners get another ten years to destroy somewhere else?* Still, it was at least a temporary reprieve for the divide. That was about a week before I arrived in the North Fork Valley, and the subject on the minds of many folks I talked with. *Where would an exchange allow them to drill? What would be the future of the vacated leases?* It seemed an ominous relief.

On the drive back to Flagstaff, my mind is full of the valley—beautiful farms, vibrant art community, busy dedicated people, the mountains, the flowers. At home, I keep track:

September brings big changes to the valley's conservation groups. Sarah Sauter, who has been an organizational dynamo for the Conservation Center, accepts the position of program manager for Oregon's Rogue River Watershed Council. But the hole from her departure is

The North Fork Valley in winter.
Photo by Rita Clagett.

quickly filled. That VISTA intern who left five hundred local residents spellbound with his Google Earth virtual flyover of the valley at the first public meeting about the BLM's lease plans back in January 2012—that was Alex Johnson. And now Alex Johnson takes over the reins of the Western Slope Conservation Center.

Two months later, another valley star is recognized when The Wilderness Society, a leading national conservation organization, hires away CHC executive director Jim Ramey to be the new outreach coordinator for its energy and climate campaign. But here also, the conservation cooperative is not left flat-footed. Board member and international trade attorney Natasha Léger agrees to lend her considerable talents as interim director.

The West Slope gets an early holiday present on December 16 of 2015 when the BLM proposes canceling leases in the White River National Forest. If they are canceled, none will be exchanged. But the following June, it releases the long-awaited draft of the Resource Management Plan revision, and although the North Fork Alternative is included, the agency's preferred plan largely maintains the status quo of fossil fuel exploitation. The Uncompahgre district manages more than 650,000 acres of surface area plus another 971,220 subsurface acres of federal mineral estate. Their preferred plan proposes to open *95 percent* of those BLM-managed lands and minerals to oil and gas leasing.

CHC and other groups, residents, and the county request an extension of the comment period on the draft RMP. BLM grants a sixty-day extension, to November 1. On that date, on behalf of CHC, Center for Biological Diversity, Earthjustice, Sierra Club, WildEarth Guardians, and Wilderness Workshop, WELC files extensive comments. It requests inclusion of a "no leasing alternative" on the basis of (1) incompatibility between fossil fuels and North Fork's other incomparable resources, and (2) the fact that current climate science and carbon budgeting demand a different set of priorities on public lands, stressing that we must keep fossil fuel resources in the ground if we hope to maintain a livable planet. By the November 1 deadline, the UFO field office has received an unprecedented 53,000 comments, including more than 42,000 recommending a No Lease Alternative. The bureau's final Resource Management Plan was scheduled for spring of 2019.

July 2016, Pete Kolbenschlag and a delegation from Colorado Farm and Food Alliance, the Western Slope Conservation Center, and Solar Energy International visit offices of both Colorado senators, the EPA, BLM, and the White House Council on Environmental Quality. Their message is that with orchards and coal having grown up together, and together powering the economy of the North Fork Valley, that heritage can transition to the twenty-first century with development of clean and sustainable energy.

Besides the West Slope's tremendous solar potential, streams and canals that drop into the valley from high in the Rockies can power small and medium-scale hydro projects, and methane emitted even from shuttered mines can be and is being captured to generate electricity. It is important, one delegate says, that the officials and agencies understand that solutions are available right now to boost the economy, create jobs, and protect the valley without contributing further to climate change. It's not jobs *versus* the environment. It's jobs, local economy, *and* protection of the environment.

Seeds of Perseverance

November 9, 2016, Donald J. Trump, who denies the fact of climate change and supports maximizing coal as well as oil and gas development, is elected President of the United States. One week later, Interior Secretary Sally Jewell, BLM Director Neil Kornze, and Governor John

Hickenlooper meet at the Colorado Capitol to announce the cancellation of twenty-five undeveloped oil and gas leases in the Thompson Divide after the BLM acknowledged deficiencies in the original analysis of sixty-five leases in the area. Jewell declares the decision "a testament to the ability of individuals, businesses, governments, and organizations in Colorado to work together to find solutions . . . for the local community, economy, and environment."

But industry representatives call it a purely political "taking." They contend that the decision came about because in its waning days, the Obama administration is getting orders from environmental groups. Texas-based oil and gas company SG Interest's Robbie Guinn and industry representatives say they will look for relief in the courts and expect the new Republican administration to uphold the old lease contracts.

Carbondale-based Thompson Divide Coalition argues that rather than being political, the decision is a response to more than 50,000 comments from local citizens and governments and those who work and recreate in the beautiful Thompson Divide. It was a decision based on democracy in action.

Glenwood Springs, Colorado, *Post Independent* newspaper (November 28, 2016) runs an online article about the cancellation of the leases, and notes that SG Interest's Robbie Guinn accused the Obama administration of colluding with environmentalists. Pete Kolbenschlag posts a comment on the irony of SGI claiming collusion, when just a few years earlier SGI itself had been found guilty of colluding with Gunnison Energy to keep bidding prices down. Rather than bidding against each other and raising prices, SGI would bid at the lowest price, two dollars per acre, and sell Gunnison half the leases for that rock-bottom rate. This practice saves considerable investment for SGI, but seriously disadvantages the government and the American taxpayer. Kolbenschlag writes, "Yes, these two companies, owned by billionaires, thought it was appropriate to pad their portfolios at the expense of . . . hard-working American [taxpayers]."

January 20, 2017, Donald Trump is inaugurated the 45th President of the United States. Almost immediately, all mention of climate change disappears from the White House website. President Obama's Climate Action Plan—a proposal to cut carbon pollution, prepare for the effects of climate change, and lead international efforts in addressing climate issues—is replaced by the incoming administration's America First Energy

Plan. "America First" calls for eliminating "burdensome regulations" that discourage development. The US Department of Agriculture and Environmental Protection Agency (EPA) employees are instructed not to communicate with the public. The EPA is ordered to remove its climate change web page. It goes back up briefly, but is removed again for "updating."

Under the obscure Congressional Review Act, which fast-tracks congressional ability to overturn rules without public input, Congress kills a rule restricting the dumping of toxic mining waste into streams. This rule primarily affects mining practices associated with mountaintop removal. Appalachian Voices, a North Carolina–based nonprofit organization advocating for "a just economy and healthy environment," says that since the 1990s coal companies have buried over *two thousand* miles of streams with heavy-metal waste, polluting water sources and causing widespread and serious health effects.

Another bill moves forward to overturn the BLM methane and waste-prevention rule via the Congressional Review Act. The America First Energy Plan states that the Trump administration "will embrace the shale and gas revolution to bring jobs and prosperity to millions of Americans." The bill fails initially, but the BLM reviews the rule pursuant to Trump's executive orders and suspends compliance deadlines. A short time later, Ruth Welch, BLM State Director, along with state directors from Alaska and New Mexico, are reassigned to non-BLM agencies, presumably for being progressive on the environment and potentially problematic to Trump's America First energy vision.

In February 2017, Houston-based oil and gas company SG Interests sues Pete Kolbenschlag for libel. In what is often called a SLAPP suit (SLAPP being the acronym for "strategic lawsuit against public participation"), SGI claims that Pete has harmed its reputation in his online comment on the irony of SGI claiming collusion against its lease interests when the company itself had been found guilty of colluding to defraud the public.

It is noteworthy that such suits are sufficiently common to have earned an acronym. Discouraging public participation is not something customarily associated with a democratic society, yet that is precisely the goal of powerful groups that feel their rights to profit should not be threatened. In her book *Democracy in Chains* (2017), Nancy MacLean recounts how, shaken by the 1954 Supreme Court decision *Brown v. Board of Education*, which ended state-sponsored school segregation, economist James Buchanan devised a step-by-step campaign to divide the public and counter popularly organized movements in order to "protect capitalism from democracy," to quote MacLean (p. 81). Buchanan and others feared that if the popular movements were allowed to influence social opinion or legislation, "overinvestment in the public sector" (social programs such as Social Security, Medicare, or Medicaid) would follow, threatening the property rights (personal profits) and power of the elite white minority.

Nearly two centuries ago, in the late 1820s and '30s, US Senator and one-time Vice President John C. Calhoun strategized ways that the wealthy minority could fight taxation and wield power in orders of magnitude beyond their numbers. Calhoun's action plans were adopted by Buchanan in an attempt to preserve elite power in the wake of integration. By 1971, lawyer and future Supreme Court Justice Lewis F. Powell Jr. looked with alarm at widespread popular uprisings such as those against the Vietnam War, along with President Lyndon B. Johnson's Great Society movement that gave rise to environmental, consumer, and occupational protections enacted in Richard Nixon's administration. Powell wrote a memo saying that the "American economic system is under broad attack." He stressed the necessity of national coordination and cooperation of business in order to gain and use political power. That quiet, disciplined movement has progressed steadily over the past half-century, enthusiastically supported and expanded by the deep pockets of Charles Koch. SLAPP, a lawsuit against public participation, would seem to be a logical outgrowth of the Buchanan, Powell, Koch campaign.

Kolbenschlag's post was essentially accurate, but he must still defend himself in court—merely the cost of doing business for billionaire companies, but a considerable financial challenge for a public-interest activist. The point of such a suit is to intimidate anyone considering crossing a powerful industry. But it just makes Pete more determined.

Though found innocent of any actionable offense and awarded attorney fees for the "frivolous" and "vexatious" complaint against him, Pete two years later must still defend himself from the company's appeal. But, as Pete says, "Free speech is protected in our republic for a good reason. Citizen input is grist for the mill of representative democracy." He now is supporting Colorado's HB 1324, which enumerates protections for journalists and other citizens to hold government accountable without fear of retaliation from the powerful.

As those of the property and profit school push their ideas of economic liberty notwithstanding the will or needs of the majority, people in the North Fork Valley as well as nationwide are responding to save what they care about, be it personal health, equality, protection of the environment, or democracy itself. As once again North Fork Valley's future seems less than secure, I wonder what buoys the community's spirit during the frequent upsets, and what inspires their terrific turnout for community events. What I often see and tend to expect is intense response briefly sustained, as people get too busy, get discouraged, burn out. But somehow, that doesn't seem to be happening here.

Much of the explanation is likely in the efforts of Citizens for a Healthy Community. CHC Acting Executive Director Natasha Léger appears to be a major source herself. Her apparent endless supply of energy seems to radiate from her crown of raven curls as she works virtually nonstop, researching, testifying, planning, informing. When I ask how she explains the community's perseverance, she reminds me of citizen science projects, like the backpack air-monitoring, that keep people active and invested. Actually taking part in gathering information, like writing letters and testifying, gives people agency—knowing they can have input in directing their lives.

Valley residents have also connected with Public Lab, an open network of community organizers that teaches inexpensive ways to investigate environmental concerns. Public Lab is a resource for people with questions about exploring water, air, or land quality, or mapping projects. The CHC website encourages locals who would like to know "how to test your ground or surface water without having to go broke, or how to see what's going on out of sight, or how to make sense of strange new odors . . . or if you have technical questions around specific tools, calibration, and installation," to ask Public Lab for help. And many do.

Citizens for a Healthy Community is primarily focused on education, keeping the community current on valley threats and successful campaigns through its website, newsletter, and public meetings, notifying the public about pending legislation and BLM plans, and giving regular action alerts. It presents public papers such as an economic analysis comparing the tax and job-creating attributes of oil and gas development with the costs to health, lost tourism, destroyed farms, and departing residents. With the second version of this detailed analysis, CHC prepared a guidebook to share with others wanting to understand the economic effects of such fossil fuel projects. This analysis is terrifically important in countering the omnipresent belief that denial of oil and gas expansion would preclude public economic health, showing that for every dollar added to the community by oil and gas development, two dollars would be lost from harm to existing uses. It appears that oil and gas expansion contributes to economic health primarily of the oil and gas industry, though of course a job is a job, and a lost job is a hardship.

For entertainment, information, and inspiration, CHC sponsors a regular movie night. Attendees watch films such as *The Age of Consequences* (2016), which looks at impacts of climate change through the lens of national and global stability, and *Tomorrow* (2019), a documentary showing positive, effective initiatives from ten countries with solutions to environmental and social challenges in agriculture, energy, economy, education, and governance. Hope is a great motivator.

CHC adds diversity and energy by partnering on projects. To help people consider the importance of protecting the valley from industrialization, in the summer of 2018, artistic center Elsewhere Studios used a $25,000 grant from the Arts in Society program to support artists for the project *INSPIRED: Art at Work*. Artists collaborated with Citizens for a Healthy Community, Solar Energy International, North Fork Valley Creative Coalition, Western Colorado Conservation Center, and the Farm and Food Alliance, along with scientists and policy makers, to "create socially-engaged artworks that address issues stemming from the impacts of legacy coal mining, such as: preservation of culture and environment, creation of a resilient economy in rural Colorado, pressures created by oil and gas development, and loss of jobs." These collaborations and a final symposium were designed to engage the broader community and promote dialogue about concerns vital to the future of this rural area.

Through WELC, CHC also partners with numerous conservation groups within the valley, throughout the state of Colorado, and far beyond. All of those groups recognize the strength gained in numbers and diversity of viewpoint. Awareness of support is motivation to persevere, a fact that the general populace of the valley seems also to have internalized.

Gathering

Another way the scattered populace remains cohesive and energized is through social events. Solar Energy International's outreach information to potential interns says of the North Fork Valley, "for a small community, we sure like to party!" Something seems *always* to be going on in the valley. Festivals, art series, farm and wine tours, Kids Pasta Project, farm to school programs, outdoor adventures from rafting to climbing, hiking, mountain biking, hunting, camping, and fishing.

Agritourism can have 30,000 visitors a year, tourists who sample wine, fresh produce, fine foods, and microbrews while admiring the area's fine art. Spring through summer sees events such as Sheep Camp Dog Trials, Pioneer Days, Cherry Days, West Elk Wine Trail, Farm to Fiddle Festival, and weekly free concerts at Pickin' in the Park.

In late July, Paonia celebrates nearly three decades of hosting the BMW Motorcycle Club of Colorado's Top o' the Rockies Rally, when more than five hundred riders congregate for three days of family-friendly fun at Paonia Park. It's all hands on deck in the North Fork Valley, with the community providing food, venues, and housing as well as taking advantage, if they wish, of an open invitation to the rally events. Each year, the valley receives a generous payback in support of local festivals and facilities, such as the BMW Club's welcome contributions to the teen center.

Dancing together in the beer garden to the BMW-rally band, listening to the concerns of a neighbor at a community meeting, cooking together at the Edesia community commercial kitchen, folks leave feeling like friends, even if they didn't come that way. Countless studies show physical and emotional health benefits from social engagement. While living in a rural community can be isolating, in this community, innumerable occasions draw people together. And that snowballs. People get acquainted; they discover that superficial differences don't preclude similar concerns; they care; they know someone cares about them. They get involved; they see they can make a difference; they feel both a reward and a responsibility.

It appears to me that conflicts I had worried might disrupt community strength—between coal miners and artists, newcomers and old-timers, perhaps between the well-heeled and the minimalists—are, if not nonexistent, at least not an issue. Of course, each person comes with his or her own personal perspective, but when they attend community meetings and write protest letters, mostly they speak for what they have in common. They speak for the valley. People chose the valley or it chose them. They know the valley and will fight to protect it. Tom Stevens, the Hellecksons, the Ela family, Mark Waltermire, and Katie Dean—these people and so many like them really care. Natasha says that this may be the most passionate and involved community she has ever known.

Beyond the networking and festivals that already bolstered the community feeling, perhaps what really cemented it was that January meeting in the Paonia Junior High gym. Five hundred people found schools and residences on maps on the wall, and again, in the virtual flyover. And they realized, *this is our home. This is a special place. We're all threatened.*

Citizens of the valley know their land like diminishingly few in modern fast-paced electronic-tied times. They farm it, fish it, run their animals on it, hike it, paint it. We can only truly love what we know. And to love is to be able to passionately protect. Like in a bucket brigade at a fire or a sandbag line fortifying a dike, you don't check each other's politics before grabbing what's being handed you. Red or blue, rich or poor, they work together to protect the valley.

Until major changes are made in laws and attitudes, the valley will continue to face challenges. But the citizens of the North Fork Valley are not fighting as a whim or the cause *du jour*. They are fighting for

Natasha on ecoflight over the North Fork Valley.
Photo courtesy of Natasha Léger.

Natasha Léger, Citizens for a Healthy Community.
Photo courtesy of Natasha Léger.

the land and water and air. They are fighting for what they love. They are fighting for their homes, for their children's future in the valley. And they know that, as Pete said at the Climate Colorado Conference, "Together, we're powerful."

Borrowed from Our Children 3

The Public Trust Doctrine

HE GYRATES ACROSS THE STAGE, arm pumping, waist-length black hair swirling, dancing, rapping to the insistent beat of the message, *listen to the music of life.* This is Xiuhtezcatl (shoo-TEHZ-caht) Martinez, fourteen-year-old Indigenous hip-hop artist. Kelsey Juliana, three years his senior, introduced him as a "rock star," and for good reason. Xiuhtezcatl gave his first performance at age 6 and, along with his younger brother Itzcuauhtli, has for years been mesmerizing audiences across the nation and the world with his social justice and eco rap.

However, rather than the energy of hip-wiggling booty-shaking Generation Z, today's audience is university professors, graduate students, practicing attorneys, and gray-haired townspeople old enough to be Xiuhtezcatl's grandparents. Still, the attendees are no less, though differently, engaged. Rhythmically nodding heads bend to take notes; toes tap and smiles spread throughout the auditorium. It's early March of 2014, and Xiuhtezcatl is a featured presenter at the thirty-third annual Public Interest Environmental Law Conference at the University of Oregon in Eugene. Between original environment-inspired songs, he reviews the context that got him to this stage.

Raised in Boulder, Colorado, in the Aztec tradition, Xiuhtezcatl had always understood that humans were put on earth to be caretakers of the land. But even as a very young child, he looked around and wondered. How could he explain trash on the stream banks, plastic bags tangled in plants? Why would workers spray poisons in the public park?

It seemed to him that people "were doing the opposite of caretaking." Then, at age 6, he saw the documentary *An Inconvenient Truth* (2006), which showed wreckage to the planet and its environmental systems being caused by excessive emissions of greenhouse gases—and that those emissions were the result of decisions based on exploitation and profit, rather than on the continued health of our planetary home. Xiuhtezcatl decided it was up to him to act, to speak, and to sing and dance for Mother Earth.

He and his friends pulled plastic off of plants, fished trash from the streams, collected what could be recycled. When he was nine, he gathered other young people and trooped to City Council to ask councilors to stop the spraying of poisons where children played. The council listened. Spraying in parks stopped, and the children felt empowered.

Xiuhtezcatl wrote lyrics about living responsibly, about social justice and environmental activism. When rapping in public with his brother, he realized that other young people—kids who would never have listened to a speech about climate change—loved the music and got the message. He formed a youth division of Earth Guardians, young activists, artists, and musicians training leaders and supporting each other to defend the planet, now with chapters worldwide. He spoke at the United Nations Sustainable Summit in Rio de Janeiro and the General Assembly at the UN in New York, and he served on President Obama's youth council. He now speaks and performs at two to three events each week as he tries to maintain other more typical teenage activities, like school. He spreads the word that "every choice we make is for or against our future."

At the same time that six-year-old Xiuhtezcatl was watching Al Gore's documentary in Colorado, twelve-year-old Alec Loorz was watching it in California. Stunned, he immediately watched it again. His generation was first in line to be hammered by climate change! He made up his mind to do something about it.

After a year of trying, Loorz was frustrated that he couldn't find any organizations inclined to take seriously either this thirteen-year-old or his goal—to end the climate crisis in his lifetime. So he formed his own organization that he called "Kids vs. Global Warming." He brought a message of hope and empowerment to kids in kindergarten through college, incorporating animation and age-appropriate science. Alec had already given more than thirty presentations when Al Gore invited him to train with Climate Project 2008, and at age 14 he became their youngest presenter.

When Alec Loorz taught kids about climate change and inspired them to live "as if our future matters," he felt he was speaking for many more than himself, finding words for what countless others were feeling. Word spread, and speaking engagements came in so fast that his mother quit her work to help him keep up.

By 2010, the nonprofit organization Kids vs. Global Warming had formed an activist branch, iMatter. Members were encouraged to "hold leaders accountable, start or join teams at school, build a national movement of young people standing together to let the world know that we matter."

Like Xiuhtezcatl Martinez and Alec Loorz, attorney Julia Olson had watched Al Gore's astonishing documentary when she escaped August's heat one day into a cool and dark theater in Eugene. Eight months pregnant with her second son, Olson "cried through the whole film." Having for fifteen years represented grassroots conservation groups—protecting rivers, wildlife, organic agriculture, and human health—she had long known the basic facts of climate change, but watching the havoc to the planet unfolding on the screen as she sat in the dark theater, her soon-to-be-born son cradled within her, she was devastated. Feeling the weight of threat to the futures of both her sons, she began focusing her work on climate change. *I've got to do this*, she thought. *I've got to do it for them.*

⁂

In 2010, Olson met Mary Christina Wood, director of the Environmental and Natural Resources Law Program at the University of Oregon School of Law. Wood had long espoused the Public Trust Doctrine and originated the idea of atmospheric trust litigation to hold governments accountable for reducing carbon pollution within their jurisdictions. These ideas intrigued Olson as a constitutional means to curb the causes and combat the ravaging effects of climate change on her children's future.

The Public Trust Doctrine traces its roots back to sixth-century Rome, when Emperor Justinian declared, "These things are by natural law common to mankind—the air, running water, the sea, and consequently, the shores of the sea." The idea being that the government holds in trust for all its citizens the resources they need to survive, and it can be held accountable if it fails to protect those resources for future generations. The doctrine was later adopted into European law, and eventually made its way from England to America. In the 1970s, some states added public trust protection to their constitutions, but, though it usually is upheld at least in matters concerning navigable waters, the federal government has never explicitly codified the doctrine.

Mary Wood had come gradually to the remedial potential of the public trust theory. Raised on the Washington side of the Columbia River, she saw signs early in her childhood of the long history of tribal life in the area, becoming aware that non-Indians were newcomers there. The Native people had maintained healthy fish populations for 10,000 years. Under the inexperienced management of the white emigrants in the Columbia River basin (as elsewhere), salmon, so abundant then, began a slide to the greatly diminished runs of today, which were markedly exacerbated by mid-twentieth-century big-dam building and continued habitat destruction. At the same time, the newcomers polluted rivers, demolished forests, filled in wetlands, and paved over agricultural land.

As an adult beginning her career teaching environmental law and Indian management of salmon, Wood thought of the maxim, "We do not inherit the Earth from our ancestors; we borrow it from our children" as an expression of a trust, in both a legal and moral sense.

So why didn't American environmental laws protect any of those resources so ravaged by the newcomers? Wood declares, "Environmental laws have failed us all" (Wood, 2014). Written with good intentions, the agencies the laws formed to protect our resources have been so appropriated by industrial money and lobbyists that through their permitting capacity, they have become the perpetrators of the very destruction they were formed to protect against. With their charters requiring management for multiple use, along with corporate pressure, agencies treat industries, not the public, as their clients, Wood says, deciding how much pollution or extraction to allow on any particular project instead of how to ensure optimal health for the resource.

To be fair, most agency professional employees try to do the right thing, but required to manage for multiple use, are caught in the squeeze play when administration appointees or directives favor industry objectives. Rather than rewriting laws, Wood suggests, we must change the paradigm interpreting them. Laws must protect the community rather than corporate profit. They must look to the long-term health of the resource rather than focus on how much it can be polluted or otherwise damaged or where to allot its rapidly diminishing components.

Julia Olson had been following Wood's work on the Public Trust Doctrine when Wood told her about Filipino attorney Antonio Oposa, who successfully represented forty-three children in the early 1990s to defend the remaining Philippine old-growth forests. In 1988, only 800,000 hectares of virgin forest remained (one hectare equals a one-hundred-meter square, or nearly two and a half acres), yet licenses had been given to log more than 3.9 *million* forest hectares. A self-described street-brawler and storyteller, Oposa feared such extensive logging would be the end of the old forests as well as most of the young ones, and decided to fight. Deeply influenced by the work of Georgetown Professor and Past President of the American Society of International Law Edith Brown Weiss's concept of intergenerational equity under international environmental law, it made sense to him that saving resources for the future should involve kids.

The Philippine lower court dismissed the case, saying that the children lacked standing, which, legally, means that to bring suit in court, a plaintiff must prove correctable injury that is sustained or imminent. Oposa, seeing the potential for grave future harm, appealed the decision to the Philippine Supreme Court, where the dismissal was reversed. The Supreme Court proclaimed it the "duty of the State to promote and protect the right of people to a balanced and healthy ecology in accord with the rhythm and harmony of nature." One judge lauded Oposa as the first to bring a case to court not only on behalf of the present generation but also future generations. Intergenerational equity became known as the "Oposa Doctrine" in Philippine and global law.

Inspired, hopeful, and passionate for the cause, Julia Olson would use the Public Trust Doctrine, Mary Wood's idea of atmospheric trust litigation, and Oposa's work in intergenerational equity. It was her intention that the courts establish government's responsibility to protect for future generations the air that all earth creatures breathe and the atmosphere that makes earth habitable. Additionally, governments would be required to secure a science-based national remedial policy to return the atmosphere to a safe balance. Olson felt the weight of responsibility to make a safe and healthy world for her children, and she wanted governments at all levels to assume that same responsibility. She argued that those things necessary for life—the air to breathe, water to drink, fertile soil to grow healthy plants, a climate comfortable to live in, and an environment to feed their souls as well as their fellow earthlings—all are a sacred trust in the care of the adults of the world for the benefit of our progeny. It should be a legal trust as well.

Our society seems to understand that concept when it comes to monetary matters. Financial trusts are common. People generally want to leave an inheritance for their children. Money. Things. Perhaps a house. Governments accept their public-trust responsibility to protect navigable waters in order to maintain commerce. Our society honors the economy. But our children could, if necessary, find other means of exchange beyond the monetary. They would be hard-pressed to find ways to live in a chaotic overheated climate with polluted air, water, and soil.

In 2010, Olson founded Our Children's Trust. By Mothers' Day 2011, Our Children's Trust had prepared to file climate lawsuits or petitions for rulemaking in fifty states plus one federal suit, all to ensure the present and future health, happiness, and lives of the youth.

As Alec Loorz's activist group iMatter was organizing a 2011 global event called the iMatter March (which would involve more than two hundred communities in forty-five countries), James Hansen, former director of NASA's Goddard Institute for Space Studies and currently Columbia University's Earth Institute Director of Climate Science, introduced Alec to Julia Olson. Olson wanted to coordinate the 2011 legal actions with demonstrations and other public events by and for kids. Excited at the idea of dovetailing the iMatter March with legal actions for youth, Alec immediately signed on. And when he understood that Our Children's Trust would be challenging state and federal governments' historic and ongoing abetting of the greenhouse-gas-producing fossil fuel industry, he wanted also to be a plaintiff for a federal case.

The Science

Greenhouse gases (GHG), described in *An Inconvenient Truth* as a major cause of climate change, are so called because they act on the planet as an unvented greenhouse does on your prize orchids, allowing the sun's rays to warm the interior and holding in the heat—not allowing that solar energy to escape. They make earth's atmospheric blanket denser, reducing the planet's heat radiation to space beyond. The balance is crucial. If less heat energy escapes than enters, the world—as would a greenhouse—warms.

Earth's atmosphere consists mostly of nitrogen, oxygen, and argon, gases that do not contribute to the greenhouse effect. The primary greenhouse gases in the atmosphere are water vapor, carbon dioxide, methane, nitrous oxide, and ozone. GHGs are not a bad thing. In the right proportions, they are not only acceptable, they are *essential* for comfortable lives. Without them, earth's average temperature would be around 0 degrees Fahrenheit, rather than its present 59 degrees. But human activities since the beginning of the Industrial Revolution have caused a 40 percent increase in the atmospheric concentration of carbon dioxide (CO_2), raising it from 280 parts per million in 1750 to 400 ppm in 2015, already seriously affecting the atmospheric energy balance. The greenhouse needs venting.

Primary sources of CO_2 emissions are the burning of fossil fuels in transportation, in heating and cooling, and in heating of kilns for cement manufacture (as well as through the calcination of the heated limestone for making cement). Less well-known sources are deforestation and soil erosion. Through photosynthesis, forests remove significant amounts of CO_2 from the atmosphere and store the carbon in their wood. When the trees are removed, atmospheric CO_2 increases partly because the trees are no longer using it, but they also release stored carbon as CO_2, primarily as the unused slash decomposes. Major emissions occur when forests are burned. More than thirty million acres of tropical rainforest are lost each year, most of it burned in order to plant crops or to run livestock. Annual CO_2 released from cutting and burning averages 1.5 billion tons, more than from all of the cars and trucks per year on all of the world's roads.

Soil erosion accounts for up to 16 percent of the amount of CO_2 released by fossil fuels, as carbon-storing organic matter is washed away and soil microbe ecosystems are destroyed. Deforestation and fire set up the conditions for soil erosion, multiplying CO_2 release. On the other hand, healthy stable soils store more than double the carbon stored in the atmosphere and about as much as is stored in the oceans.

Primary sources of methane emission are animal agriculture in confined animal feeding operations (CAFOs), natural gas distribution, drilling, flaming and leaks, and, importantly, landfills. Nitrous oxide is released in fossil fuel burning and fertilizer use.

If worldwide emissions remain steady, scientists estimate that global warming could pass 2 degrees centigrade above preindustrial levels by 2036, a threshold the UN's Intergovernmental Panel on Climate Change (IPCC) earlier called the upper limit to avoid dangerous effects on ecosystems, biodiversity, and livelihoods the world over. Commissioned by the 2016 Paris talks, an October 2018 IPCC report compares that earlier 2 degree warming limit to 1.5 degrees Celsius, finding a stunning difference in water stress, food scarcity, habitat loss, sea-level rise, and heat-related deaths. The clear mandate is an urgent effort toward a ceiling of 1.5 degree Celsius rise.

Toward a New Paradigm

Our Children's Trust has joined numerous other environmental and justice groups to insist on an immediate reduction in fossil fuel use, coupled with vigorous advancement of alternative energy sources. The urgency is hard for leaders of relatively unaffected countries to come to grips with. The climate is not so bad now. We've had some pretty major storms, but we've had storms in the past. What's the panic?

Once greenhouse gases are in the atmosphere, they hang around for a while. Much of the carbon dioxide, which makes up nearly a quarter of greenhouse gases, stays in the atmosphere from twenty to two hundred years, but a portion remains for several hundreds to thousands of years. The wide range is because of the different processes removing carbon dioxide from the atmosphere. Up to 80 percent is photosynthesized by plants and dissolved in the ocean within twenty to two hundred years, but some processes, such as chemical weathering and rock formation, can take from hundreds to thousands of years. Methane's short-term role as a GHG is nearly ninety times more powerful than that of carbon dioxide. It has a briefer atmospheric lifespan and is a far smaller percentage of total atmospheric greenhouse gases. Still, while it is present, it carries a wallop. Ensuring its percentage doesn't increase is critically important.

Also not readily apparent just looking out the window are the climate-change feedback loops—warming effects that become causes for more warming, which can be both difficult and frightening to quantify. Bright ice, for instance, reflects sunlight back to space, but melting ice exposes dark soil and dark sea that absorb heat and increase warming, which of course melts more ice. Melting permafrost—permanently frozen soil—would not only contribute to sea-level rise, but also expose peat beds that would emit that potent GHG, methane. As much as seventy billion tons of methane could be released from the world's largest peat bog in Western Siberia, causing more warming to melt more Arctic permafrost. Rainforest drying and increased forest fires are among other effects that could accelerate climate change. Also, for each degree earth's temperature rises, methane entering the atmosphere from wetlands and lake sediment will increase several times, presumably because of faster decomposition of organisms in a warmer environment. These considerations—the relative strengths and atmospheric lifespans of greenhouse gases, and the various feedback loops—increase the urgency for immediate reduction in greenhouse gas emissions.

Earth Institute Director and former NASA scientist James Hansen says that if we had begun in 2002 reducing CO_2 emissions by 3.5 percent a year, we could have decreased atmospheric CO_2 to 350 ppm, considered the safe maximum, by the end of the century. Beginning in 2015 would have required a 6 percent reduction per year, a formidable goal. Not beginning to act until 2030 would leave the CO_2 levels above 350 ppm for 700 years. Hansen says that considering feedback loops and nations' reluctance to reduce carbon emissions, the world is approaching tipping points—rapid change with effects that would be irreversible—if nations don't slow greenhouse gas emissions directly. The October 2018 IPCC report advised that in order to keep warming under 1.5 degrees Celsius, global greenhouse gas emissions must be reduced by *60 percent* by 2030. So few years! The challenge is daunting, and will take all hands on deck including those of government and industry, but it can still happen if we tune in to the kids.

Getting those two heavy-hitters—government and industry—on board is exactly Julia Olson's goal. Our Children's Trust (OCT) began filing cases May 4, 2011. *Alec L* (for Alec Loorz) *v. McCarthy* (Gina McCarthy, then Environmental Protection Agency administrator) was filed in US District Court for the Northern District of California. Alec joined four other youth plaintiffs to seek an immediate turnaround in national energy policy in order to decrease greenhouse gas emissions by 6 percent beginning in 2013. The plaintiffs were represented pro bono by OCT's partner law firm of Pete McCloskey, co-founder of Earth Day and former Republican US congressman. The five teens plus two nonprofit plaintiffs, Kids vs. Global Warming and WildEarth Guardians, sued the heads of the EPA and of the Departments of Commerce, Interior, Defense, Energy, and Agriculture. Thirteen international experts on climate science, energy, and US policy supported the legal filing.

In April 2012, several California construction and manufacturing associations received permission to intervene in *Alec L v. McCarthy*, and immediately moved to have the case dismissed. They argued that a "small group of individuals and environmental organizations" shouldn't "dictate the economic, energy, and environmental policies of the entire nation." A spokesman added that granting the plaintiffs' demands "would carry serious and immediate consequences" for productivity and cost, and would decrease global competitiveness. He pointed out that the plaintiffs' requests "could substantially eliminate the use of conventional energy in this country," which of course is one of the major aims of the suit. The intervenors also insisted that the plaintiffs hadn't proven they had a legal right to sue.

This was the first climate-change lawsuit to request a national plan and focus on risks to future generations and to national security. As such, it was transferred to Washington, DC, because of its nationwide significance and for the convenience of the federal government. May 2012, four industry-hired law professors filed amicus curiae (friend-of-the-court) briefs arguing that there is no federal public trust, and if there is, the emissions of greenhouse gases by the public (driving trucks and cars, running industries with smokestacks) is the sort of public interest the Public Trust Doctrine seeks to protect. In other words, noted Julia Olson, they're saying that the Public Trust Doctrine's purpose is to allow pollution. The highest and best use of the atmosphere is for a garbage dump.

On May 31, the District Court for the District of Columbia issued a decision granting the government defendants' and fossil fuel intervenors' motions to dismiss the youth plaintiffs' case.

After expressing pride in the young plaintiffs and disagreement and disappointment in the decision, Julia Olson said, "We agree with [district court] Judge Wilkins that 'this case is about the fundamental nature of our government and our constitutional system.' That system mandates protection of our fundamental right to a healthy atmosphere on which humanity depends."

A frustrated Alec Loorz, now 19, said, "The court wants us to find . . . common ground with the fossil fuel industry and government, but I believe that as long as these institutions value profits and power over the survival of my generation, there can be no common ground. . . . [W]e will not give up. We will continue fighting to protect our planet for our generation . . . and all who follow, for as long as it takes."

Our Children's Trust petitioned to appeal to the Supreme Court, but their petition was denied. Olson told her sons, ages 6 and 9, "Don't worry. Mom's going to take care of this."

As the Alec L case was being argued in Washington, DC, legal actions were active in every state in the nation. Many were dismissed with government questioning jurisdiction (it's up to the legislature, not the courts), standing (the kids haven't been hurt yet), the use of the Public Trust Doctrine (the atmosphere isn't part of the Public Trust), or claiming plaintiffs were asking the federal government to set international law (other nations pollute too). Across the board, governments moved to dismiss the cases, but sometimes the court denied their motions. Washington and Oregon had cases running on parallel tracks:

May 4, 2011	Andrea Rodgers and Richard Smith file in Seattle against Washington State Governor's Office, Department of Ecology, Public Land and Fisheries and Wildlife Chris Winter and Tanya Sanerib of Portland's Crag Law Center file in Eugene on behalf of Oregon plaintiffs.
January 2012	State of Oregon moves to dismiss. State of Washington acknowledges dire effects on human and natural systems if emissions are not controlled, then moves to dismiss.
February	Washington Trial Court says the case is obviously very important, then grants the government's motion to dismiss.
April	Washington's Andrea Rodgers appeals the dismissal, stating that in spite of the state's recognition of impending catastrophe, it has taken no action. She stresses that this case gives the judicial branch, which is charged with interpreting and applying the Public Trust Doctrine, an opportunity to address its obligations to protect present and future generations.

Dr. James Hansen supports Rodgers's petition for appeal, stressing the need for immediate action without which humanity and the rest of nature would experience "calamitous consequences" as tipping points are reached and points of no return are crossed. Failure to act quickly would "consign our children and their progeny to a very different planet, one far less conducive to their survival."

The state moves to strike Dr. Hansen's declaration, but Washington's Supreme Court denies the motion. In June, numerous faith-based groups file amicus briefs supporting the youths' petition for review. Referencing the declaration of international human rights and basic morality, they urge Washington to honor its role as a trustee of the atmosphere, a "fundamental issue of broad public importance, meriting direct and accelerated review."

April 2012	Lane County Oregon's Judge Rasmussen sides with State of Oregon motion to dismiss.
July	Washington youth file briefs, noting that courts in Texas and New Mexico had issued rulings that the atmosphere is indeed a public trust, at least in those states.
August	Greg Costello of Western Environmental Law Center submits amicus briefs by top law professors around the country, supporting both the Washington plaintiffs' petition for review and the atmospheric trust claim.
September	Washington State files opposition brief denying government public-trust responsibility to protect the atmosphere or to take climate action.
October	Oregon's Court of Appeals agrees to hear the plaintiffs' (Kelsey Juliana, then 16, and Olivia Chernaik, 12) appeal of Judge Rasmussen's dismissal of their right to be heard and present evidence to compel the state to create a viable climate recovery plan.

Oregon youth's attorneys Chris Winter and Tanya Sanerib file the Juliana and Chernaik opening brief in December, supported with amici briefs from political leaders, legal scholars, agriculture, Native student groups, and businesses. The amici cite ocean acidification, decrease in agricultural productivity, decrease in snowpack and water supply as climate change effects of particular concern. The following year, October 2013, the Court of Appeals announces that the Oregon Atmospheric Trust's dismissal appeal would be argued at the University of Oregon School of Law January of 2014, the year and venue we first found Xiuhtezcatl Martinez rapping his penetrating message and rhythms across the stage.

꩜

January 16, 2014. A hundred or more kids, their families, local leaders, and community members troop into the law school auditorium to hear whether the Our Children's Trust atmosphere case against the State of Oregon can move forward. US District Court Judge Ann Aiken questions both sides including several questions concerning a recent Pennsylvania case, relevant because Oregon and Pennsylvania constitutions have identical language on reserved rights for the people. The constitutions state, "All power is inherent in the people, and all free governments are founded on their authority and instituted for their peace, safety, and happiness." The Pennsylvania court struck down a pro-fracking law and, rejecting a defense brought by the Commonwealth, found it was within the job of the judiciary to determine whether the laws of the state require or prohibit certain acts of other branches of government. Pennsylvania's Supreme Court also described the people's public trust rights as inherent and inviolate and part of a social contract with the people.

Oregon's attorney general agrees that the atmosphere should be considered part of the public trust, but argues for the state that the court should leave rulemaking up to the legislature. Unfortunately, so far, the legislature hasn't acted, and protecting the atmosphere is a race against time. Kelsey and Olivia's attorney, Tanya Sanerib, recalling that the judiciary had to step in to require civil rights era legislation, says, "The job of our courts [is] to protect and preserve the rights of the next

The earth is in our hands.
Photo courtesy of Robin Loznak/Our Children's Trust.

generation." Once again, both sides await the court's decision. Will she uphold Judge Rasmussen's decision to dismiss or will she allow the young people's case to proceed?

Nearly three years have passed since the first climate cases were filed. Kids are three years older. At an average of twenty-nine billion metric tons (a metric ton equals 1,000 kilograms or about 2,204 pounds) a year, humans have belched eighty-seven billion more metric tons of carbon dioxide into the air since these cases began. The CO_2 in the atmosphere has risen from 390.0 parts per million in 2011 to 397.2 ppm in 2014. The countervailing forces hold fast to their respective goals: government stonewalling, industries wanting to continue business as usual, both facing bright and passionate kids and their supporters, all wanting to save their futures. But increasingly, some members of the judiciary are beginning to envision a new way forward.

On June 11, 2014, the Oregon Court of Appeals rules that a trial court must hear the Juliana and Chernaik case against the state of Oregon, reversing Judge Rasmussen's decision to dismiss. The trial court must decide whether the atmosphere is a public trust that the State of Oregon has a duty to protect. This win comes on Kelsey Juliana's last day at South Eugene High School. Immediately after graduation, she will join the Great March across the nation, whose organizers hope will inspire and motivate residents, officials, and the media along the route to address the climate crisis. Seven million steps for the environment, the organizers say. Nearly four hundred people from thirty-seven states and seven countries officially walk all or parts of the event. Hundreds more join as the group surges through their towns. Beginning in Los Angeles in March, the group arrives 3,000 miles away in the nation's capital in November. Still in school when the march sets out, Kelsey joins in Nebraska and walks fifteen to twenty miles a day, comparing notes with other activists and meeting a stimulating cross section of citizens along the way. "I have everything to gain from taking action," she says, "and everything to lose from not."

April 8, 2015, nearly four years from the original filing, hours before court is to convene, crowds of people wearing rain hats or carrying umbrellas queue up in a steady drizzle outside the federal courthouse in Eugene. Some huddle under the roof or pack the stairways, singing, chanting, waving signs, cheering the throngs of kids, signs hoisted, parading from their schools or spilling out of buses. More than four hundred individuals from across Oregon—many of them students—flood the courtroom and the overflow room or are part of a vigil outside the courthouse, all eager to hear Juliana and Chernaik request the right to demand effective climate action from the state government. The luck of the draw determines that the case will once again be argued before Judge Rasmussen.

Arguing the case for the plaintiffs—the kids wanting livable futures—Chris Winter says, "In the face of unprecedented and irreversible harm to our natural resources, this case gives Oregon the chance to lead our nation in protecting our atmosphere and all natural resources so our children and grandchildren enjoy a viable life on Earth." The state's attorneys renounce any such obligation, contending that the Public Trust Doctrine does not apply to the atmosphere. And once again, they move to dismiss the case, a motion that will be accepted or denied by the judge in the next months.

After the two-hour trial, Kelsey Juliana, who flew home to Eugene from North Carolina's Warren Wilson College, says she is grateful for her attorneys but "disappointed and confused why my state is continuing to battle and resist" protecting the environment for present and future generations.

Kelsey of course is not the only one disappointed and confused. What an education this must be for these young people. Any youthful idealism about an all-knowing and compassionate State is summarily dashed. But imagine 8- or 11- or 14-year-olds learning that though their future is precarious, they have the possibility to make a difference. How empowering and exciting. They learn about the creaking machinery of the judicial system, the humanness of judges, the profit motive of industries and therefore of some officials. But they also learn about the dedication, determination, passion, selflessness, empathy, and compassion of numerous individuals, professionals, and groups. The youths are finding models and paths for their lives.

May 11, 2015. Without considering any of the undisputed evidence presented in court, Lane County Circuit Court Judge Rasmussen agrees with the defendants that the case should be dismissed. He "question[s] whether the atmosphere is a natural resource at all," and writes, "It is difficult for the court to imagine how the atmosphere can be entirely alienated."

Julia Olson says, "By ignoring the factual record, including expert testimony on the catastrophic impacts facing Oregon's public trust resources, the court did these youths and all Oregonians a great injustice today. The decision is unsupported by law and by fact, and we have every confidence in the appellate process." Our Children's Trust will once again appeal Judge Rasmussen's decision to dismiss.

Meanwhile, in Seattle

While Olson is preparing to appeal in Oregon, less than three hundred miles north Our Children's Trust attorneys have been in and out of court for their *Zoe & Stella Foster v. Washington Department of Ecology* case and rebuffed frustratingly often. In August of 2014, Washington's Department of Ecology responds to the most recent suit, saying, "I appreciate the concern and desire of the youth petitioners to reduce greenhouse gas emissions for improving the environment for themselves and future generations. However, Ecology denies your petition for rulemaking in favor of its current approach to reducing greenhouse gas emissions."

Arguing that the "current approach" is inadequate for climate recovery, the eight Washington youths file a challenge in King County's Superior Court to Department of Ecology's denial of their rulemaking request. Attorney Andrea Rodgers argues that Ecology's position is "reckless," as it is the "youth petitioners' generation whose lives will be altered by the present generation's failure to take action."

June 23, 2015, King County Superior Court Judge Hollis Hill sides with the Washington youth plaintiffs. Judge Hill orders Ecology to reconsider the petition asking for CO_2 reductions and to report back to the court in two weeks as to whether Ecology will consider the undisputed science necessary for climate recovery.

The effect of this decision is that "for the first time in the United States, a court of law has ordered a state agency to consider the most current and best available science when deciding to regulate carbon dioxide emissions," Andrea Rodgers says.

"Kids understand the threats climate change will have on our future," says thirteen-year-old petitioner Zoe Foster. "I'm not going to sit idly by and watch my government do nothing. We don't have time to waste. I'm pushing my government to take real action on climate."

August 25. In spite of the King County Superior Court's June 23 order for the state to reconsider its first denial of the youths' petition, the Department of Ecology, which says it is working on its own version of a rule—"doubling down on obsolete science"—as Rodgers puts it, denies the petition for a second time. The Washington youths file a response to the King County Superior Court.

November 3, 2015, Washington's young petitioners find themselves once again in Judge Hollis Hill's King County Superior Courtroom listening to their attorney's impassioned arguments. They are shocked to hear the defense state, "I do not know if there is an inherent right for a healthful environment" and to tell the court that it is up to the legislature to decide.

But on November 19, Judge Hill declares that the youths' "very survival depends upon the will of their elders to act now, decisively and unequivocally, to stem the tide of global warming . . . before doing so becomes first too costly and then too late." The judge determines that the state has a "mandatory duty" to "preserve, protect, and enhance the air quality for the current and future generations." The youths and their attorneys are thrilled and hopeful.

However, the following February 26, Washington's Department of Ecology withdraws its proposed rule to reduce carbon emissions in the state. April 6, 2016, youth petitioners ask the court to step in once again and April 29, Judge Hill orders Ecology to enact an emissions reduction rule by the end of 2016 and make recommendations to the state legislature on science-based greenhouse gas reduction in the 2017 legislative session.

June 1, 2016, the Inslee administration releases a totally inadequate Clean Air Rule, decreasing carbon by a mere 1.7 percent, "failing the children of Washington," says Julia Olson. July 25, Washington youths provide the administration with climate science to show how to update the state's Clean Air Act cutting greenhouse gases enough to be effective. At this point, an annual emission decrease of *8 percent* is required, *along with carbon sequestration*, in order to reduce atmospheric carbon to 350 ppm by the end of the century, the maximum level scientists feel is allowable for a livable climate.

November 22, youth petitioners return to court for a hearing in which Governor Jay Inslee's administration must show why it is not in contempt of the court's prior order to protect the constitutional rights of the youths from climate pollution. December 19, 2016, Judge Hill rules that the trial for the youths suing the Department of Ecology can again move forward, this time with a constitutional claim that adds Governor Jay Inslee and the State of Washington as defendants.

"In this global race against time, we must think bigger than what Governor Inslee believes is politically possible," said Julia Olson. "We must strive for what science requires. Reality and our children demand no less, for in the end, there is no negotiating with nature."

Government Culpability

On International Youth Day, August 12, 2015, as state trials go forward in Washington and Oregon—along with Colorado, Massachusetts, Pennsylvania, and elsewhere—twenty-one young people, ages 8 to 19, file a federal suit. Inspired by the Alec L case, Xiuhtezcatl Martinez had connected with Julia Olson and Our Children's Trust in 2011, and later encouraged Olson to prepare this case against the US government. Martinez, now the youth director of Earth Guardians, returns from his third address to the United Nations just two months before the case is filed.

With the assistance of Our Children's Trust, young people from across the nation request that President Obama immediately institute a national plan for decreasing the atmospheric concentration of CO_2, now at 400 parts per million, to 350 ppm by the end of the century.

Our Future. Youth plaintiffs are right of center. Julia Olson in front, in dark suit and white collar.
Photo courtesy of Robin Loznak/Our Children's Trust.

Besides stressing the urgency for the courts to step in when executive and legislative branches continue "such intense violations" of the plaintiffs' constitutional rights, OCT's Olson points out that the federal government has known about the dangers of excess greenhouse gas in the atmosphere for more than fifty years, but has continued to encourage fossil fuel use nonetheless.

As early as the mid-1950s, the director of Scripps Institute of Oceanography and the US Office of Naval Research both had noted the link between increased carbon dioxide in the atmosphere from the burning of fossil fuels and higher temperature, as well as more frequent hurricanes. In 1965, President Lyndon Johnson released a report predicting the melt of the Antarctic ice sheet, and the rise and acidification of the oceans. It stated that burning coal, oil, and gas at the 1960's rate of six billion tons a year would add 25 percent more carbon dioxide to the atmosphere by 2000. Fifty years later, the CO_2 concentration approaches 400 ppm, 25 percent above the 1965 levels. Though the report contained dire warnings of climate effects that would eventually be uncontrollable, the government continued to promote and subsidize development of fossil fuels.

Similar reports from the Atomic Energy Commission and the National Academy of Sciences continued through the 1970s. In the mid-1980s, Congress held hearings on the greenhouse effect and examined implications for public policy. A bipartisan subcommittee requested that the EPA study potential health and environmental effects, a report the agency produced in detail. In 1988, James Hansen testified to the Senate, stressing effects of climate change and the urgency of action. The 1990s and early 2000s saw reports to Congress that said "limiting emissions cannot wait" and that the burden of doing nothing will fall on "generations of people who are not even born." Still, awash with decades of information and warnings, the government chose inaction.

In a 2017 letter to the editor in the Eugene, Oregon, *Register-Guard* newspaper, lead plaintiff Kelsey Juliana, with nine co-signers, writes that it may not be apparent how the US government can be a cause of climate change. So she lists a sampling of ways, including insufficient fuel-efficiency standards; leasing land to corporations for oil, gas, or coal mining; giving tax breaks and subsidies to fossil fuel companies; and failing to regulate greenhouse gas emissions of the federal energy system. She says, "When you add up all these actions, the US government, more than anyone else, is responsible for the level of carbon dioxide pollution that will determine the climate in our lifetimes."

Therefore, as Julia Olson points out, the government not only bears the sovereign responsibility, as the trustee of essential resources, to protect a livable future for the children, it must also assume partial culpability for the climate's current untenable trajectory.

❧

November 2015, three months after Our Children's Trust files the federal case, major fossil-fuel trade groups ask permission from the US District Court for the District of Oregon to join the US government as co-defendants in the young people's climate case. Those groups include the American Petroleum Institute, representing 625 oil and natural gas companies; the National Association of Manufacturers, representing over 11,000 large and small manufacturing companies throughout the nation; and the American Fuel and Petrochemical Manufacturers, representing ExxonMobil, BP, Shell, and Koch Industries. Calling the children's case a direct threat to their businesses, the trade groups explain that having to reduce greenhouse gas emissions would decrease sales of the product they have spent time and money developing. Immediately on receiving Judge Coffin's permission to intervene, the industry representatives, along with the government, move to have the case dismissed.

"Seeing giant fossil fuel corporations inject themselves into this case, which is about our future, really demonstrates the problem we are trying to fix," said plaintiff Xiuhtezcatl Martinez. "The federal government has been making decisions in the best interest of multinational corporations and their profits, but not in the best interest of my generation and those to come. Instead of changing their business model to meet the scientific reality of climate change, these companies are demanding we adapt to an uninhabitable world that supports their profits. When you compare the two, I think it's clear that our right to clean air and a healthy atmosphere is more important than their 'need' to make money off destroying our future."

By now, the kids are getting used to the government and corporate's reflexive request to dismiss their case. Several plaintiffs and their attorneys have suggested that the defendants are afraid to have the public hear the facts. But once again, they will have to await the judge's decision. Will the defendants' motion to dismiss prevail? Or will the youths' case go forward?

❧

Early the next year, 2016, the federal court case gathers friends. In mid-January, the Global Catholic Climate Movement (GCCM), an international network of over 250 Catholic organizations and thousands of individuals, and the Leadership Conference of Women Religious (LCWR), representing more than 40,000 "women religious" across the nation and world, file amici curiae briefs. The amici state, "As people of faith and as citizens of the United States, we are deeply concerned about the policies, plans, and practices of the federal government, which do far too little to achieve the reduction in fossil fuel emissions necessary to ensure the health and well-being of our children and our planet home, now and into the future."

In September of the same year, the League of Women Voters of the United States and of the state of Oregon offer assistance. The League believes that averting the damaging effects of climate change requires action both from individuals and from governments at the local, state, national, and international levels. The amici note that climate change is already causing environmental damage that will have significant economic and social impacts. Though the League of Women Voters usually focuses its efforts on the legislative and executive branches as it works to protect rights, particularly those of underrepresented citizens, it recognizes the necessity of judicial involvement when, as in this case, the other branches have failed their citizens.

March 9, 2016. Hundreds converge on the federal courthouse in Eugene to hear whether the case that Bill McKibben and Naomi Klein call "the most important lawsuit on the planet right now" will be allowed to go forward. Under a twenty-five-foot banner proclaiming "KIDS SUING TO SAVE THE CLIMATE," elementary to teenage students, and adults their parents' and grandparents' ages, jostle on the courthouse steps amidst signs, posters, chants, speeches, songs, and a pervasive feeling of hope. The lines waiting for seating open for students, for the infirm, and for the Raging Grannies, singing elders in colorful clothes and wildlife-habitat bonnets. The courtroom fills, with hundreds still in line. Arguments stream via video feed to three additional courtrooms plus another in Portland.

During the two-hour hearing in this David-and-Goliath case, the attorneys for the defendants (the federal government and the world's largest energy companies) urge US Magistrate Judge Tom Coffin to dismiss the youths' lawsuit. As before, they question the kids' standing, contend they have not proven injury, say that such regulations should be decided by legislatures and agencies. Attorneys for the oil and gas companies add that a finding for the plaintiffs would cause their clients considerable harm and that the industries are already constrained by the Clean Air Act and other environmental regulations.

The youth plaintiffs argue that the government is violating their constitutional rights to life, liberty, and property by allowing dangerous levels of greenhouse gases to be released into the atmosphere, and by actively promoting the development and use of fossil fuels. They also seek a court order requiring the government to create a plan that will decrease GHG emissions sufficiently to avoid climate catastrophe.

A month later, April 10, 2016, Judge Coffin's decision is in, shaking the foundations of fossil fuel companies worldwide. The case can go forward! Kids scream, laugh, bump fists, and high-five. Julia Olson takes a deep breath and tries to stem premature happy tears.

In deciding against the federal government and fossil fuel industry's plea to throw out Our Children's Trust climate case, Judge Coffin says, "the unprecedented lawsuit [addresses] government action and inaction [causing] carbon pollution of the atmosphere, climate destabilization and ocean acidification . . . [T]he intractability of the debates before Congress and state legislatures and the alleged valuing of short term economic interest despite the cost to human life, necessitates . . . the courts to evaluate the constitutional parameters of the action or inaction taken by the government. This is especially true when such harms have an alleged disparate impact on a discrete class of society." The discrete class, of course, represented by the young plaintiffs, not yet (when the case was begun) of an age to vote, and therefore, without voice or representation.

Celebrating the decision to move forward, Xiuhtezcatl Martinez says, "When those in power stand alongside the very industries that threaten the future of my generation instead of standing with the people, it is a reminder that they are not our leaders. The real leaders are the twenty youths standing with me in court to demand justice for my generation and justice for all youth. We will not be silent. We will not go unnoticed, and we are ready to stand to protect everything our 'leaders' have failed to fight for. They are afraid of the power we have to create change. And this change we are creating will go down in history."

The next step for the federal case is a review of Judge Coffin's decision by Judge Ann Aiken of the same court.

Youth plaintiffs celebrate Judge Coffin's decision that their case may move forward.
Photo courtesy of Robin Loznak/Our Children's Trust.

The behind-the-scenes work of the attorneys and expert witnesses is almost nonstop as they research, coordinate, and prepare briefs. But further behind the scenes are other support team members. They work with kids and the media, and they do office work, field organizing, and global program management. The lawyers work pro bono, so financial support needs to appear from somewhere. OCT has an active Board of Directors and receives support from numerous other organizations, such as 350 Eugene, which is instrumental in organizing events and the art that makes events zing, and Environmental Law Alliance Worldwide, a global alliance of attorneys, scientists, and others who support grassroots work toward justice and sustainability worldwide.

350 Eugene's Joanie Kleban, the magician behind the banners, signs, and flags at rallies, has a long history of working for the greater good. She came to Eugene from Massachusetts and Antioch College in Ohio in 1976 and decided to stay, after meeting her future life partner, Cary Thompson. Together Kleban and Thompson worked distributing organic produce and exploring fair trade and human rights, as Thompson helped set up Eugene Sister Cities in Russia, Nepal, Japan, and South Korea. Kleban's interests in art, fair trade, human rights, and community-building then motivated more than two decades running

the Greater Goods import store, which helped support more than two hundred craftspeople in about thirty countries. And now Kleban's talents direct art projects that involve myriad people and give the messages extra pizzazz when they appear on the media:

> *There is no planet B.*
> *The atmosphere is a public trust.*
> *The fierce urgency of now.*
> *Let the youth be heard.*

❧

September 13, 2016, twenty-one hope-filled kids troop into Judge Aiken's court. It goes well. The judge is attentive and asks perceptive questions. Fossil fuel companies continue their indignation, and the government continues to question judicial authority to arbitrate.

On Wednesday November 9, Donald Trump, an avowed denier who once contended that climate change was a hoax perpetrated by the Chinese to gain an unfair trade advantage, is declared the next President of the United States. Children, attorneys, environmental groups, and fossil fuel companies ponder the consequences. The following day, Judge Aiken rejects the fossil fuel industry and the government's motion to scuttle the children's case, ruling that the twenty-one young plaintiffs, ages 9 to 20, may move their case to trial.

"This action is of a different order than the typical environmental case," writes Judge Aiken. "It alleges that defendants' actions and inactions—whether or not they violate any specific statutory duty—have so profoundly damaged our home planet that they threaten plaintiffs' fundamental constitutional rights to life and liberty . . . I have no doubt that the right to a climate system capable of sustaining human life is fundamental to a free and ordered society . . . Federal courts too often have been cautious and overly deferential in the arena of environmental law and the world has suffered for it."

"It's clear Judge Aiken gets what's at stake for us," says seventeen-year-old plaintiff Vic Barrett, from White Plains, New York. "Our planet and our generation don't have time to waste. If we continue on our current path, my school in Manhattan will be under water in fifty years. We are moving to trial and I'm looking forward to having the world see the incredible power my generation holds. We are going to put our nation on a science-based path toward climate stabilization."

"This is a critical step toward solution of the climate problem, and none too soon as climate change is accelerating," says Dr. James Hansen, world-renowned climate scientist and guardian in the case for all future generations. "Now we must ask the court to require the government to reduce fossil fuel emissions at a rate consistent with the science."

Government Admissions

January 13, 2017. A week before Donald Trump's inauguration, lawyers for the outgoing Obama administration concede many of the plaintiffs' allegations in *Juliana et al v. United States*. They acknowledge that the use of fossil fuels is a major source of greenhouse gas emissions, "placing our nation on an increasingly costly, insecure and environmentally danger- ous path." And they admit to a half-century of institutional awareness of research about the effects of fossil fuel emissions on the atmosphere, including long-lasting and worsening changes to the climate. These stunning admissions are now part of the record that will follow the case forward.

A week and a half later, January 24, Our Children's Trust asks the Trump government to retain all records relating to climate change, energy, and emissions, in addition to conversations between the gov- ernment and the fossil fuel industry. The naming of defendants in the federal suit changes to specify President Trump and his administration. Asked how it feels to be suing President Trump rather than President Obama, a young plaintiff says it feels good. "Even if President Obama wasn't doing enough, at least he was trying." And now the Trump ad- ministration orders a review of the Clean Power Plan, the legislation the industry claimed was constraining them, along with numerous environ- mental rules designed to protect the public from polluted air and water. At the same time, the administration intends to open public lands and the ocean to oil drilling and encourage a renewed push to mine coal.

March 7, the Trump administration files two motions, one to appeal Judge Aiken's order to let the case go forward, the other, a request to delay case preparation until after the appeal is considered. They then ask that review of both motions be pushed ahead, complaining of the burden imposed by the plaintiffs' January 24 request that defendants keep their records. March 10, fossil fuel industries join the government's request to appeal.

"We have a classic example of the government's misplaced pri- orities," says OCT's Olson. "They prefer to minimize their procedural obligations to not destroy government documents over the urgency of not destroying our climate system for our youth plaintiffs and all future generations."

Twenty-year-old plaintiff Alex Loznak, a student at Columbia University, says, "This request for appeal is an attempt to cover up the federal government's long-running collusion with the fossil fuel industry. My generation cannot wait for the truth to be revealed. These documents must be uncovered with all deliberate speed so that our trial can force federal action on climate change."

OCT in DC

One of the goals of Our Children's Trust is to gain public attention to the threats of climate change to future quality of life. Youth presence at popular events furthers that goal. Earth Day 2017, the twenty-one young plaintiffs arrive in Washington, DC, for a week of marches and rallies. April 22 is the March for Science, a unique event, as scientists ordinarily prefer to be in their labs doing science rather than walking in the street carrying signs. But this march is not for a specific cause—it's not for or against nuclear power or GMOs—this is about the very idea of science, science that has been pushed aside, ignored, and defamed by the new administration. The importance of respecting science is key to the kids' climate cases. Hundreds of thousands come out around the world. Among them are at least hundreds of journalists.

Australian Broadcasting Corporation's Conor Duffy caught up with some of the Climate Kids. He wondered how they had become involved in a court case against the federal government.

Curly-haired nine-year-old Levi Draheim's home is on a barrier island off Florida's coast, a five-minute walk from the beach. He says that at the current rate, his home could be under water before he is his dad's age.

Nathan Baring, 17, is a high school junior in Fairbanks, Alaska, 120 miles south of the Arctic Circle. The Arctic is warming two to three times as fast as the rest of the planet. In the past, snows could be expected before the end of the summer. Now the ground often remains dark until late fall. At this rate, Arctic sea ice will be no more than a note in the history books by mid-century. With average annual temperatures up to 4 degrees Fahrenheit higher than normal, undreamed-of changes to forest, ocean, and human ecosystems await the next generation.

"I am a winter person," Baring says. "I won't sit idly by and watch winter disappear."

Amy Goodman from *Democracy Now!* asks Julia Olson and some of her young clients about their landmark suit. Olson notes that while the Obama administration admitted that the kids are facing a crisis, the current administration is working hand in hand with the fossil fuel industry to fight the kids.

Oregon's twelve-year-old Hazel Van Ummersen tells Goodman that "it's extremely important for us young people to stand up to our government, where adults are doing nothing to prevent climate change and to stop the harmful effects of ocean acidification and sea-level rising."

Asked what she thinks is getting in the way, Van Ummersen says that our president "does not believe that science is real . . . but we have science to prove him wrong. We will see him in court and we will win."

The week winds up with the April 29 March for Climate, Jobs, and Justice, and more speeches and interviews. From the steps of the US Supreme Court, Xiuhtezcatl Martinez says, "For the last several decades, we have been neglecting the fact that this is the only planet we have and that the main stakeholders in this issue [of climate change] are the younger generation. Not only are the youth going to be inheriting every problem that we see in the world today after our politicians have been long gone, but our voices have been neglected from the conversation."

May 1, 2017, with the kids back at their various homes throughout the nation after an adrenaline-pumping week in the nation's capital, the chances increase that the youth voices will be heard. In Eugene, Oregon, Judge Thomas Coffin recommends denial of the Trump administration and fossil fuel defendants' motions to derail the youths' climate case with an early appeal, and denies the motion to put the trial on hold until such an early appeal is decided. Though the defendants have often contended that this is not a question for the judicial branch, Judge Coffin says, "The court may make findings that define the contours of plaintiffs' constitutional rights to life and a habitable atmosphere and climate, declare levels of atmospheric CO_2 [that] will violate their rights, determine whether certain government actions in the past and now have and are contributing to or causing the constitutional harm to plaintiffs, and direct the federal defendants to prepare and implement a national plan [that] would stabilize the climate system and remedy the violation of plaintiffs rights." Judge Coffin sets the court date for February 5, 2018, the case to be heard by Judge Ann Aiken.

Julia Olson says, "It's time for the defendants to accept they are going to trial and not try to continue bending the rule of law to delay a judgment in this case. President Trump must accept that the courts do not do his bidding, and in a court of law, 'alternative facts' are considered perjury."

Government Stonewalling

Meanwhile, the Trump administration continues its tactics to delay trial. The government's request for "interlocutory appeal" (appeal of a ruling before all parties' claims are resolved) is denied by Judge Aiken. Next, the defendants petition the Ninth Circuit Court of Appeals in San Francisco for a "writ of mandamus," which asks the higher court to tell a lower official, in this case justices Coffin and Aiken of the district court, to correct an error. What the defendants consider an error is the "discovery" petition—the plaintiffs' request that the government maintain all relevant documents. They also call for a stay of the proceedings.

The potential to delay is built into judicial proceedings. Hasty judgment or difficulty appealing a questionable judgment wouldn't insure fairness. However, moneyed individuals or corporations can take advantage of such opportunities and drag out proceedings to bankrupt their opponent or simply stall a potential undesired judgment. For a big corporation or other well-financed body, it's a cost of doing business, and deductible. In the government's case, years of delay are financed by the taxpayers. For private individuals, payment for dragged-out cases comes out of their pockets.

The government's pleas will be considered before the Ninth Circuit Court late in 2017, making the kids' February 5, 2018, court date unlikely.

❧

The youth plaintiffs arrive in San Francisco for the December 11, 2017, trial before the Ninth Circuit Court, partly with the help of crowd funding. They are buoyed not only by that support, but also by eight friend-of-the-court briefs representing dozens of religious, legal, and environmental organizations. Twenty-year-old Jacob Lebel says, "It's an amazing, heartwarming feeling to have the support of all these accomplished organizations and individuals and to know that they have our backs in this fight. Right now, my home in Oregon is choking in smoke while the Southwest drowns in floodwaters. It makes me hopeful that communities of faith and global advocacy groups are standing united to defend youth's right to safe climate and I look forward to moving rapidly to trial."

This writ of mandamus case is argued before a three-judge panel, each of whom rigorously question both sides. February 5 (the date set by Judge Coffin for the kids' case against the government finally to proceed) passes with no court case and no word, but is celebrated by the plaintiffs and friends posing with posters and banners proclaiming, "#RatherBeInCourt!"

But while the kids await word on the Ninth Court's decision and a hoped-for new court date, wins are stacking up in other states and around the world. In early 2018, the Colorado Court of Appeals supports the plaintiffs in a case begun in 2013 by Xiuhtezcatl Martinez, his brother, Itzcuahtli, and four other teens. The youth requested that the Colorado Oil and Gas Conservation Commission deny any drilling

Levi Draheim protesting yet another government stonewall.
Photo by Leigh-Ann Draheim.

requests unless proven by the best available science and confirmed by a third party that drilling could be done in a manner (1) protecting Colorado's atmosphere, water, wildlife, and land resources; (2) not impacting human health; and (3) not contributing to climate change. In the past, industry and the commission operated under the premise that health and environmental interests could be "balanced" against economic interests. Requiring health and the environment to take precedence over private profits is a remarkable and welcome change.

Positive outcomes are also pending in Alaska, Maine, Massachusetts, New Mexico, North Carolina, Oregon, and Washington. All fifty states have prior or developing action. Meanwhile, Norway has adopted a public-trust-based constitutional climate amendment; the Urgenda Foundation, a group representing multiple generations, won a victory in The Hague District Court, requiring the Netherlands to reduce greenhouse gases 25 percent by 2020; and the Pakistan Supreme Court says that seven-year-old Rabab Ali's constitutional climate case may move forward.

Kids Rise

Good news notwithstanding, the courts move slowly and the climate won't stop changing while plaintiffs wait. But the kids aren't waiting either. Youth groups are forming and active all over the nation and the world. EG 350 is a climate group begun by students at Oregon's South Eugene High School, combining the goals of Earth Guardians with those of 350.org.

Earth Guardians is the global youth-driven climate education group begun by Xiuhtezcatl Martinez. It urges kids to reconnect with and stand up for their environment. The number 350 refers to the safe parts-per-million level of atmospheric carbon dioxide, and to the international group 350.org, begun in 2008 by Bill McKibben and a group of university friends to build a global grassroots movement for the climate.

Probably the world's largest climate action organization, 350.org is active in at least 188 countries where it works toward the reduction of atmospheric carbon dioxide, now well over 400 ppm, to a safe level (350 ppm) by the end of the century. The means are education, activism, and lobbying for legislative action to stop all new fossil fuel exploration. With the mantra, "Keep it in the ground!" 350 fights any new coal, oil, or gas project, as it also fights for universal clean energy.

The goal is to build a zero-carbon economy—and a just and equitable world, built on the power of ordinary people.

A group of South Eugene's EG 350 members agreed to share their lunchtime with me in an interior garden at their school. I asked how their group had formed and how they had come to activism. In 2015, Corina MacWilliams, a senior in spring 2018, was moved by the urgency of the cause and the energy of the people when she and her aunt marched for the climate through downtown Eugene, their homemade signs held aloft. MacWilliams and some others wanted to capture that momentum and to communicate to their generation information gleaned on the march, along with the vitality that comes from working together. The students realized their high school had no standing climate club, and so EG 350 was born.

Senior Stella Drapkin said becoming an activist was a natural outgrowth of living in this green college town, bordered by rivers and with a view of mountains. "Plus, several of the plaintiffs on the kids' climate case are friends of mine."

"You don't have to be a plaintiff to be a climate activist," South Eugene junior Ian Curtis added. His urge began when he realized that he'd been taking for granted the natural beauty that surrounded him. He wanted instead to pay attention to the forests and rivers and mountains and to protect it all. At the same time, he was frustrated by political unresponsiveness to climate change. The subject seemed always to get pushed aside. "I started activism in a small way," Ian said, "and discovered it was fun and I could make a real contribution."

Serena Orsinger, then a junior at South Eugene, smiled at the memory as she traced her earliest interest to pictures she saw at age 3 of polar bears balancing on ever-receding floes of ice. "I was terribly worried about those poor bears and immediately wanted to do something about their plight," she said. She has been an activist for the environment ever since, and from age 14, volunteering on projects such as building schools from recycled materials in the Dominican Republic and spearheading funding for ceramic water filters to counter waterborne diseases.

Junior Tyee Maddox Atkin had to miss our lunchtime get-together, but explained later in an email that he was born into a love and appreciation for the Pacific Northwest and into deep environmental activism. His mother was a founder and is current associate director of the Environmental Law Alliance Worldwide (ELAW), an international network of environmental lawyers and scientists. Tyee lived in and volunteered with that culture, appreciating life in a bilingual household when his family was years-long host to a Peruvian colleague, and discovering disparities in environmental justice as his mom's international colleagues discussed environmental issues in their home countries.

Tyee's father grew up on a farm near southern Oregon's Illinois River and worked his way through college, living in a little self-built covered wagon. With a passion for preserving the natural world, he now practices public interest law, representing nonprofit groups dedicated to working for a healthy planet. Influenced by all of those things, Tyee is viscerally attuned to environmental injustice and the need for western cultures to live more lightly.

Corina MacWilliams tells me that EG 350's more than fifty members have ongoing projects within the areas of education, activism, and civics. The club got South Eugene High School certified as an Oregon Green School, earning a $500 prize that it used to replace cafeteria plastic with reusable silverware. The group secured funding to replace asbestos-lined single-paned windows with energy-saving double-paned, and created and distributed a recycling guide. EG 350 designs presentations, panels, and performances to pass on basic information about climate change, describing principles, feedback loops, and tipping points, and stressing the urgency of action. The club organizes marches in support of Our Children's Trust or other environmental causes, preparing with participatory art projects for posters to carry or inform. "And kids know little about civics these days," MacWilliams notes, "so we try to remedy that as well."

MacWilliams, Serena Orsinger, and other members of EG 350 attend City Council meetings, urging councilors to push forward the goals of Eugene's Community Climate and Energy Action Plan created in 2010. That goal was for the city to be carbon neutral by 2020, with the updated Climate Recovery Plan requiring reduction of fossil fuel use by 50 percent of 2010 levels by 2030. So far, progress is less than encouraging. The original draft, plus its updates and ongoing encouragement, are largely the work of passionate young people pressing the point that climate work must be top priority. Nothing else matters if humans and other species cannot live in their environment. Each year— each *day*—the work is delayed, it becomes more difficult.

Ian Curtis was excited about the upcoming Oregon Youth Legislative Initiative. Representatives from EG 350, as well as from high school climate groups throughout the state, were preparing to go to the capitol for lobby day, to talk with legislators, and to demonstrate for the Clean Energy Jobs Bill. "We need to get youth involved in the political system," Ian says. "And legislators need to hear youth voices."

The bill, SB 1070, is a cap and invest bill. It cuts allowable pollution rates with the goal of keeping statewide greenhouse gas emissions to about 80 percent below 1990 levels by 2050, and commits the largest polluters, those who emit more than 25,000 tons GHG per year (equal to 133 railcars of coal annually) to buy permits, with collected fees to be reinvested in clean energy jobs.

Lobby day turned out to be the largest in the state's history for climate lobbying, according to 350 Eugene. Over five hundred climate warriors trouped through the marble halls, including at least twenty youths from eight Oregon schools. Ian Curtis says that as the students talked with their senators and dropped by the offices of a dozen representatives, they met the full range of opinions. They met senators highly supportive of the bill and even feeling that it doesn't go far enough, and others who feel government should be involved in a minimum of governance, and certainly not in climate change. The pro-bill lobbyists wanted the bill to come to vote immediately. Some legislators agreed; others wanted to wait, hoping for more bipartisan support. Curtis points out that "political leaders have been saying 'next year' for decades and we've run out of time. If the bill passed today, it would not be implemented until 2021 and climate action cannot wait." As it turned out, the bill was introduced but the legislative session adjourned with it still on the president's desk. Governor Brown and numerous legislators pushed for action in the 2019 and 2020 legislative sessions but were unable to reach a quorum either time.

Students left Salem excited to have been part of the political process, feeling their voices had been heard. Whether or not they'd changed minds on this, their first lobbying day, they agreed there would be many more in their future. These smart passionate students were just getting started.

Tomorrow

The youths from South Eugene's EG 350 club had thought a lot about how their dedication to the fight for a viable future would affect their life choices as they graduated from high school and looked toward the next chapters in their lives. Serena Orsinger noted that her generation has a powerful opportunity to be heard. Still having the memories and passion of childhood but now also with the budding intellect of maturity—and being on the brink of voting age—they are in a unique position. But as to the future, she says, "No one really *wants* to do this work. Like generations before us, we'd rather enjoy our lives doing whatever our talents lead us to do and be able to trust we'll have a livable world to do it in."

Stella Drapkin has spent her life pointing to a career in music, but, she says, "that doesn't preclude activism. I can use my music selfishly or I can use it to benefit the world." Corina MacWilliams says that knowing what they know, not working for a viable climate "would be immoral." Serena Orsinger adds that they don't really know that what they do will be sufficient, but to do nothing would have—and she hesitates—"a bad outcome. So we might as well do something." MacWilliams adds, "I think it's the most important thing in my life. I dedicate every decision that I make to mitigating climate change."

Feeling strongly the clash between an economic system built on profit and growth versus an environment and justice requiring simple living and empathy, Tyee Atkin plans a gap year after graduation to strengthen his self-sufficiency skills, improving his ability to live harmoniously with the world. He is committed to gardening, with its opportunity to learn about life-essential systems as well as grow food, and to carpentry, which gives him the skills to build his own house. He wants also to hone the lost skills of tracking, making fire, and building rudimentary shelter. But he will leave plenty of time to deepen his relationship with Oregon, rafting, backpacking, and climbing. Then he'll build a little house on wheels. "I believe that—especially in America—we need to learn to live with less," Tyee says. "Tiny homes are a great way to tread lightly on the environment." Once he has his little house, he'll head to college to study in an environmental field, always supporting groups like EG 350. He appreciates an activist group's ability to amplify voices while it reminds him "none of us is alone in this struggle."

Having been privileged to visit with such bright, aware, dedicated kids gives me tremendous hope for their generation and for the world. Our Children's Trust court cases grab public attention, alerting people to facts they might not otherwise see. Youths worldwide organize, feel empowered to lift their voices, and exercise their right to a viable future. And the kids are influencing cities and other local governments to commit to action even as the federal government lags behind.

Court Rejects Government Petition

On March 7, 2018, the Ninth Circuit Court of Appeals rejects the Trump administration's "drastic and extraordinary" petition for writ of mandamus. Judge Thomas Coffin sets the new date for *Juliana v. United*

States climate trial for October 29, 2018, to be heard before Judge Ann Aiken in Eugene, Oregon.

As the kids await their October court date, the United Church of Christ launches a 1,000 Sermon Initiative, where youths across the nation talk with their churches about climate change, and about #youthvgov. Renee Serota, a sophomore in Environmental Economics and Policy at UC Berkeley, tells the Congregational Church in Sonoma on Earth Day 2018 the importance of working together. She explains that the point of the 1,000 Sermon Initiative is to build community support within the church system, going beyond individual activity and mobilizing for political and legal action against climate change. Individual actions like recycling or conserving water feel good and do make a difference, but the far greatest amount of environmental damage comes from governments and corporations. Individuals can't have much effect against such large entities, but masses of people working together boycotting, lobbying, advocating for new laws can completely change a harmful practice or system. Serota tells the congregation that addressing climate damage requires large-scale structural change, and the only way to accomplish that is with the hard work of millions of people demanding new laws and policies, people willing to make those new laws work.

Vic Barrett, a #youthvgov plaintiff and University of Wisconsin-Madison student, talks about resilience—that capacity to be tough and to recover from difficulties. They remind that resilience is not always something striven toward. For some, it is an essential part of life. Self-describing as "Latinx, black, and queer," Vic notes that resilience may be won "by the adversities life has awarded you." And for this reason, in order to "fight the powers that hand away pieces of our environment for profit," it is necessary to enlist people from the margins of society. These are the people who know how to be resilient. Resilience has been a requirement for them "to survive in a society not made for them."

Juliana v. United States plaintiff Kiran Oommen tells the congregation of the Alki United Church of Christ in Seattle, Washington, why he is giving his life to the climate struggle. He acknowledges that with current prevalent attitudes, he doesn't know if it's a fight he can win. But he knows he needs to find purpose and he knows he needs "community to embody it." He says he's not here to win. He's here to fight. Explaining the climate case, Kiran says that as the government's support of the fossil fuel industry has aggravated climate change, it is violating the constitutional rights of youth to life, liberty, and property. Rather than

OCT Executive Director and Chief Legal Counsel, Julia Olson.
Photo courtesy of Robin Loznak/Our Children's Trust.

asking for compensation, the young people are asking for a court order for a science-based recovery plan. Kiran's activist philosophy is in learning to live despite knowing their work may not succeed. Being a part of the climate case gives him a glimpse of a "more purposeful now." What keeps him going, he says, is the community. The issues are complex and it's never easy, but by "sacrificing our lives through dedication, we are saving ourselves from a life of meaninglessness."

Julia Olson says that the climate kids will not give up until they win. Having courts recognize the atmosphere as a public trust would truly, in Naomi Klein's words, "change everything." Once the courts agree that preserving a healthful planet should be a protected goal, it will take immense numbers of people working together to make it happen—people of all ages and ethnicities. The tremendous growth of youth activism is and will continue to be critical. But one thing is clear: the law and public buy-in are in a race against time. The urgency compounds to help save this earth home that we have borrowed from our children.

Converging on the Cove 4

J ODY MCCAFFREE HEAVED an exasperated sigh. *No should mean no!*
Dead should stay dead! After devoting over a decade of her life to
battling this liquefied natural gas (LNG) project, and after it had
been rejected three separate times by a federal regulatory agency, as if
straight from a horror movie, the zombie project resurrected once again.

At issue is a Canadian company's plan to transport about 430 billion
cubic feet annually of highly flammable fracked gas in a thirty-six-inch
high-pressure pipeline, 229 miles across southern Oregon and ending
within two miles of McCaffree's front yard in Coos Bay, Oregon. This,
the proposed Pacific Connector Gas Pipeline, would be laid from Malin,
Oregon (about thirty miles south and east of Klamath Falls), cutting a
hundred-foot swath across traditional tribal lands and burial grounds and
impacting more than 600 private properties. It would continue through
public land and old-growth forests, across and through the Cascade and
Coastal mountain ranges, and across or under 485 wetlands and waterways. Its destination would be the proposed Jordan Cove LNG Energy
Export Project, a fourteen-story gas liquefaction, storage, and export
facility to be built on a sand spit in a tsunami and earthquake-prone
subduction zone in the Coos Bay estuary.

McCaffree's group, Citizens Against LNG (CALNG), is concerned
about the roughly 17,000 citizens in the towns of North Bend and
Coos Bay who live in the hazard area, if something should go wrong
with what one local referred to as a "fourteen-story bomb." McCaffree
spends countless hours researching, educating the public, and petition-
ing politicians on risks and concerns, and why they should take official
positions opposing the project. CALNG is incredulous that a project of

such a scale is even being considered so near a population center in the Cascadia fault zone. With major quakes occurring around every 240 to 300 years, the area is overdue for a big one—8.7 to 9.2 on the Richter scale, according to Oregon State University seismologist and geophysicist Chris Goldfinger—with a 40 percent chance of its occurring within the next fifty years.

CALNG is frustrated that industry would be destroying habitat not only for shore birds, but also for oyster and clam beds and for numerous fish species. There are problems for people's pocketbooks as well. The group wonders how many of the public realize that while Asian recipients of American gas would be paying less than they are accustomed to, natural gas bills would go up for Americans. It's the old law of supply and demand. We currently have an oversupply of US gas. If we ship out our excess, supply becomes more limited here and therefore costs rise. CALNG also wonders about the local impact when a thousand or more nonresident construction workers invade the area adding to or competing with the local workforce.

Jody McCaffree stumbled upon the earliest version of the Jordan Cove project in 2004, when the plan was for an import facility. There had been nothing in the press about it and no public discussion as far as she knew. But one day in May, while McCaffree's husband, Bill, was on a field trip with their junior-high-age daughter, he was told of a liquid nitrogen plant proposed for their town, all decisions having been made behind closed doors. Jody McCaffree, astounded at this report and dedicated to the idea that the people should have a say in such decisions, decided to investigate.

The first thing she discovered was that the "LN" initials that someone thought meant liquid nitrogen, were but part of the actual initials, LNG, standing for liquefied natural gas. At that time, gas shortages were predicted, and the Canadian company hoped to import gas at the Coos Bay port and sell it in California. The company applied for and received a conditional federal permit, but meanwhile, hydraulic fracturing (fracking) of shale formations produced a gas glut. In 2012, Calgary-based Veresen Co. refashioned the project as a gas export facility to ship Canadian and Midwest gas to Asia. April of 2012 the permit for import was rescinded by the Federal Energy Regulatory Commission (FERC): death number one. But that didn't discourage Jordan Cove. They would apply again, but this time for export.

Jody McCaffree had not intended to get involved in fighting Jordan Cove's LNG project. But as she researched liquefied natural gas and understood more of the potential for accidents, leaks, explosions, and fires, along with the inconceivable idea of siting such a project in an earthquake and tsunami zone, she became increasingly alarmed. She thought, however, that she could organize one public meeting to inform her fellow citizens about the facts of the proposal, and that would be the end of it.

It wasn't.

The Jordan Cove project looks like a gift from heaven to many local leaders and some underemployed folks. Coos Bay, for thousands of years a bountiful home for fishing, hunting, and gathering tribes of the Coos, Lower Umpqua, Siuslaw, and Coquille, developed a proud industrial history after European pioneers settled in. Fur traders arrived in the first flush of immigrants. In the mid-nineteenth century, the Coos Bay Commercial Company established Empire City in the middle of the Hanis Coos Indian Village, and brought in speculators and service businesses. Gold prospecting, coal mining, ship-building, sawmilling, farming, and ranching followed. In the early twentieth century, the Southern Pacific railway connected the southern coast to the Willamette Valley and work began on deepening the harbor, allowing transportation of goods both by water to San Francisco, and inland, by rail. The Port of Coos Bay became the largest deep-draft coastal harbor between San Francisco and Puget Sound. During the 1930s, shipyards built minesweepers and rescue tugs to protect the coast before and during World War II. Mid-century, national lumber giants bought over 300,000 acres of old-growth forests and set up operations there, needing employees in the mills as well as in shipping. That industry grew until the Bay's port was the largest lumber exporter in the world. But by the 1980s private forests were exhausted. Then came the recession of the early 1980s, conservationists' opposition to uncontrolled logging on public lands, automation in the mills, and whole logs shipped abroad—all resulting in jobs lost until the dreams of some seemed a mere distant memory.

Many locals long for a return to the glory days, now three decades past, and see Jordan Cove as the conduit for getting there. City and county leaders focus on the boost to their operating budgets to be derived from taxes on the estimated $9.8 billion construction project, though, as CALNG points out, Jordan Cove is requesting a fifteen-year tax exemption and suggesting much smaller enterprise-zone contributions. The four counties to be crossed by the pipeline (Coos, Douglas, Jackson, and Klamath) will also be anticipating help for their coffers, an obvious selling point. The fact that Jordan Cove could be the first, and perhaps even sole, West Coast port shipping LNG to Asia from the lower forty-eight inspires visions of wealth and international recognition, even though LNG terminals are operating on the west coast of Alaska and Mexico, and several are proposed for Canada's west coast. Still, for those focusing on industry and export and old-time stories, it's an exciting vision.

At the same time, CALNG and others challenge the idea that the Bay area is depressed and in need of corporate help. The group points to unemployment figures that are close to US average and to the growing level of tourist dollars, saying it would be far better to capitalize on what the area uniquely has going for it, and that Jordan Cove could threaten those environmentally appropriate businesses.

Jordan Cove hired locals to talk up the project. Many wanted to believe that this was the ticket to prosperity. Some truly did and do believe. McCaffree had been involved in other struggles, though, health and education campaigns, where she discovered that if you keep presenting the facts of a situation, eventually even some of the unbelievers

Popular tourist destination Shore Acres State Park, a dozen miles southwest of Coos Bay. Photo by Judy Richardson.

will come to see the truth. So in 2006, she and several others formed the Citizens Against LNG group, and the citizens grassroots network got busy learning about liquefied natural gas and the permitting process, and educating the public about the project.

CALNG points out potential dangers, damage to the environment, and economic realism: most of the advertised and temporary two thousand construction jobs would be supplied from outside the area; just over a hundred local jobs would be permanent (about the number needed for a new Walmart, one opponent noted); a current global market glut of natural gas will only get worse as companies ramp up their production to keep the pipes flowing, foretelling a bust to follow the exciting boom. If that weren't enough, the LNG project would threaten or possibly even destroy current area businesses, such as oyster farming, clam digging, and fishing as well as real estate, recreation, and tourist trades. And it would diffuse local energy and focus that could otherwise be directed into sustainable economic projects such as wind or wave energy, or sustainable fishing or crabbing industries, all more appropriate for this coastal town.

Longtime Coos Bay restaurateur and social critic Wim De Vriend sold his book called *The Job Messiahs* (2011) from behind the counter of the Blue Heron Bistro before his retirement at the end of 2014, recounts Ted Sickinger of Oregon Live (2014). In the book, he enumerates four decades' worth of taxpayer-funded "pie-in-the-sky thinking" to return the area to its former economic glories. Some of the unrealized dreams included a chromium smelter, a garbage-burning plant, a fish-waste processing plant, and oil-drilling platforms. He sees Jordan Cove as just one more pipe dream. De Vriend and others want the economy to improve, but not at the expense of the health of the people, the region, or the planet.

CALNG is only one of dozens of groups fighting the Jordan Cove Energy Project. More *Gulliver's Travels* than David and Goliath, the multimillion dollar "Zombie-Beast" Pipeline and its affiliated export facility are being tied down from all directions. But it's *Gulliver's Travels* meets the *Walking Dead*, as the pipeline and export facility refuse to stay tied down.

CALNG represents the concerned voices of the coastal communities where the export facility would be sited. But numerous other voices add urgency—Indigenous groups, landowners, climate and conservation groups. Native tribes—the Karuk, Hoopa, Yurok, and Klamath—passed resolutions against the pipeline running through traditional villages and burial grounds, wrote letters and newspaper columns, organized information meetings, joined demonstrations.

About 60 percent of the nearly seven hundred landowners along the pipeline route united to fight against the loss of large swaths of their land, and the threat of eminent domain. Losing land to a wealthy foreign company does not sit well with people who have sunk their life's savings into their properties. And many who live where their parents and their parents' grandparents have lived risk loss or damage not only to their properties but also to material and emotional history. Joining these struggles for human rights are voices for the rivers, for the land,

Proposed 229-mile Pacific Connector Pipeline route through private property, public lands, 400 waterways, two mountain ranges, and terminating at Coos Bay, Oregon.
Map by Imus Geographics.

and for wildlife. Of overarching importance are the voices for the climate, which is the conductor for the entire choir of other voices.

KS Wild

The Klamath-Siskiyou Wildlands Center (KS Wild), then primarily a public lands watchdog group with offices in Ashland, was an early organization to be aware of the LNG Project, learning about it as Jody McCaffree did, when it was still intended as an import terminal.

KS Wild's mission is to protect and restore wild nature in the Klamath-Siskiyou region of southwest Oregon and northwest California. This eleven-million-acre jumble of mountain ridges and peaks, along with its wild rivers and gorges, wooded hills and flowering meadows, earns raves from the World Wildlife Fund for its exceptional biodiversity. The International Union for the Conservation of Nature calls it one of only seven Areas of Global Biological Significance on the North American continent, and it has been proposed as a World Heritage Site and UNESCO Biosphere Reserve.

Around 200 million years ago, about the same time the Blue Mountains were forming, an island arc that would become the Klamath Mountains moved eastward, colliding repeatedly with the Pacific coastal North American plate, each time leaving bits of the old island accreted to its new North American home. Some Klamath Mountain rocks are as many as 500 million years old. Intruding into these ancient rocks are peridotite and serpentine, rocks formed under extreme pressure in the earth's interior. These rocks are deficient in calcium and potassium and high in magnesium, iron, and nickel, challenging plant growth to a limited number of tolerant species.

The Klamath Mountains' complex geology and rugged terrain, combined with precipitation patterns ranging from less than twenty inches in the eastern reaches to as much as 120 inches near the coast, account for some of the biodiversity, with species adapting to the variety of soil types as well as specific climatic conditions. That the mountains run both north and south and east and west contributes to varied climate and habitats as well.

This area escaped glaciation when most of the continent was frozen during the last ice age, its benign climate providing refuge for many species and a long habitable period for others to evolve to new conditions. Additionally, the Klamath-Siskiyou is bordered by a diversity of ecosystems: chaparral and oak woodlands, Cascadia and the Cascade Range, the Sierra Nevada, California's Central Valley, and Oregon's Coast Range. The Klamath-Siskiyou tied together microclimates and species from these varied bordering regions—the Klamath Knot, so beautifully described by David Rains Wallace in his book of the same name (1983)—preserving diverse ancient species found nowhere else.

Around 3,500 plant species make the KS their home, including thirty-six species of conifers, more than appear in any other temperate forest in the world, and a mosaic of northern California and Pacific Northwest deciduous trees as well. Nearly three hundred of the area's vascular plants are endemic. Rich and diverse animal populations find homes here also, including rare and endangered salamanders, Marbled Murrelets, and the Pacific fisher—a fierce house-cat-sized member of the weasel family. The first, and as yet only, place in Oregon to host returning wolves was in the Klamath-Siskiyou, and it was through the KS that wolves were reintroduced to California. Aquatic species are rich and varied too in this, the West Coast's most roadless area, with the largest concentration of Wild and Scenic Rivers in the lower forty-eight.

All of which makes the area particularly important to preserve, and KS Wild is and always has been on the lookout for any kind of threat. Already a frustrating amount of logging and mining has occurred, and though the area that includes the juncture of the Cascade, Klamath, and Siskiyou mountains finally obtained national monument protection in 2000, that too is being threatened. KS Wild had spent years protecting wildlands, wildlife habitat, and remaining old-growth forests when the first rumors of the Jordan Cove Energy Project and its connecting pipeline came along. This was a scheme that would clear-cut a hundred-foot path through eighty miles of public land, seriously fragmenting habitat and obliterating nearly three hundred acres of old growth. KS Wild took note and began contacting agencies and political leaders.

The Klamath-Siskiyou Mountains comprise the watershed for the Rogue River to the north and the Klamath River in the south. The region also claims the Illinois River, the Salmon, Trinity, Upper Sacramento, Smith, and Chetco Rivers and all of their tributaries. The Jordan Cove Energy Project's Pacific Connector pipeline would cross or tunnel under wetlands and waterways at 485 individual locations.

In 2008, the Klamath-Siskiyou Wildlands Center's Lesley Adams decided to spin off a separate program to look after the health of the Rogue River and its tributaries. Adams had mobilized communities to engage in local, state, and federal decisions concerning the common good and had coordinated campaigns and fought for justice through the legal system. Now, using environmental laws such as the Clean Water Act, the new program, the Rogue Riverkeeper, would work for clean water throughout the Rogue River Basin's three and a third million acres of rivers and streams that provide southern Oregonians with water for drinking, irrigating, and outdoor recreation.

Though an arm of KS Wild, the Rogue Riverkeeper would have its own autonomy and be a member of the Waterkeeper Alliance, an international network. Waterkeeper Alliance was born in 1983 on New York's Hudson River when a group of fishermen decided they'd had enough of cavalier industrialists dumping their waste in the river and destroying both quality and quantity of fish. It would be an interesting history to research. I have a feeling the fishermen didn't play nice. An online note refers to their "tough, grassroots brand of environmental activism." But however they played it, it worked. They accomplished a dramatic cleanup of the river and inspired a global organization.

The Waterkeeper Alliance networks with waterkeepers for more than three hundred waterways on six continents, each with the mantra, "Without water there can be no life, and without clean water, there can be no healthy life." The Alliance claims to be the largest and fastest-growing nonprofit focused solely on clean water. Every day polluters dump toxins into waterways, and waterkeepers fight daily for "swimmable, drinkable, fishable waters." In each organization, the manager is called the "keeper" of the given waterbody (river/lake/sound/bay). Now Lesley Adams, who launched the Rogue Riverkeeper for KS Wild, is the western US senior organizer for the Waterkeeper Alliance, and Robyn Janssen is Rogue Riverkeeper.

Rogue Riverkeeper

Robyn Janssen may have been born a water sprite. In love with wild nature from childhood, she took the earliest opportunity to work for a local southern Oregon rafting company. After an end-of-summer four-day rafting trip on the Wild and Scenic Rogue at age 18, she didn't want to leave. The operations manager was taken by her comfort on the water. He had been searching for more female guides, and he asked

Robyn if she'd be interested in training. She was not merely interested: she was thrilled. She ran her first guiding trip on the Rogue just two years out of high school. In the beginning, there were particularly macho guests who doubted the ability of this slip of a girl with her long dark curls, and Robyn rather delighted in proving her ability to them. She loves showing first-timers the river and the canyon and explaining the ages of the rocks and how the river cut through them to carve its canyon. She points out the problems too: temperatures that are too high for salmon habitat, bacteria, sedimentation, and toxic algae. Most of these threats come from construction, logging, or agricultural runoff. One of Janssen's favorite things is teaching women—especially young women—how to row, how to finesse and work with the river.

Now as Riverkeeper, Janssen shares her knowledge and appreciation for the river with her clients, with other organizations working for the environment, and with the public. Rogue Riverkeeper's mission is "to protect and restore water quality and fish populations in the Rogue Basin and adjacent coastal watersheds through enforcement, advocacy, fieldwork, and community action."

Robyn and Willy on the Wild and Scenic Rogue.
Photo courtesy of Rogue Riverkeeper.

Even before becoming Rogue Riverkeeper, Robyn Janssen pointed out problems as she took her clients down the river. Jordan Cove would bring many more—to the ecosystems, to potable water, and to the fish themselves.

The project, including the pipeline, would impact the Coos, Coquille, Umpqua, Rogue, and Klamath watersheds. About fifty feet would be leveled on each side of the pipe for construction, maintenance, monitoring, and to make it less likely that an explosion in the pipe would start a forest fire, or that a forest fire would cause an explosion in a pipe. Near waterways, that clearing would remove shade, increasing already-compromised water temperatures. In Oregon, the Rogue Basin is second only to the Columbia in wild salmon runs, but salmon require cool water. More exposed waterways also mean more evaporation in a part of the country suffering already from drought and above normal temperatures. Digging and drilling around and under rivers and through smaller streams increases sedimentation and turbidity, impairing the quality of spawning and rearing habitat for salmon and steelhead. Digging can introduce or dislodge chemicals into the water, impacting potability of municipal drinking water supplies and rural wells in addition to harming aquatic life. Maintenance of the pipeline would include herbicide spraying and fertilizing of grass planted to slow erosion, introducing more chemicals to drain into the waterways, polluting the water and stimulating toxic algae blooms.

All told, this 229-mile pipeline would travel through streams, rivers, sloughs, wetlands, and a bay, negatively impacting riparian and upland habitat, forests and farms, ancient cultural sites, residential lands, and recreation. The Rogue and Klamath Riverkeepers and the Waterkeeper Alliance request all hands on deck.

SOCAN and Rogue Climate

In September of 2012, six dozen southern Oregonians concerned about climate change and its ramifications gathered to brainstorm ways they might address these problems, both individually and collectively. Named for their mission, Southern Oregon Climate Action Now (SOCAN) grew quickly to a membership of a thousand.

SOCAN's goal is the reduction of atmospheric greenhouse gas concentrations to 350 parts per million by 2050, the level scientists feel is maximum to avoid catastrophic change. Conserving energy, switching to renewable energy, and examining typically consumptive lifestyles are all part of the equation. Reducing greenhouse gas emissions requires action by individuals, corporations, and governments locally, nationally, and worldwide, and it necessitates working together, all of which SOCAN promotes.

Seeing the need for education both for themselves and for the public on causes, consequences, and solutions for climate change, SOCAN organizes presentations, courses, and videos directed to all ages and segments of the population. A letter-writing group emerged to keep facts and solutions in the media and the public eye, and, recognizing the need for collective and government action, those letters also went to legislators. The organization also partners across the region, the state, and the nation with other groups focused on climate issues.

SOCAN's first public function, sometimes called the Salmon Event, was held in February of 2013. With a goal "to bring people together and kick-start our creativity and innovation," the event featured music and presentations, but perhaps most importantly, it featured wide public involvement in a symbolic art project. Citizens of all ages decorated over a thousand cardboard tiles inspired by two questions: "What do you love about living in the Rogue Valley?" and, "What worries you when you think about climate change?" These tiles became the scales on a 120-foot mosaic of a salmon, an icon of the valley of the Rogue River.

Having people think about and express what they loved and what they feared cemented emotional attachment to the project. And that's the key, isn't it? If we focus on what we love—know it, understand it—we are compelled to protect it, aren't we? Then as we address how what we love could be threatened by climate change, mightn't that reveal a path to its protection? The people thought deeply about the designs of their individual tiles. Contributing them with those of others to the expansive and beautiful collective art project demonstrated the magic that can happen when many hearts, minds, and hands join together: *I made a meaningful little tile. I was part of the making of a stunning 120' salmon. I have proclaimed what I love and will join all of these others to defend it.*

One of the organizers of the great Salmon Event, advertised to bring climate concerns home (with art), was Hannah Sohl. Hannah was newly back home to the Rogue Valley after graduating from Colorado College in 2011, then spending a year traveling to Bangladesh, Canada, and Russia studying grassroots organizing and the relationships between human communities, rivers, and migratory fish. This year of independent study was supported by a Thomas J. Watson fellowship, an award to "engage [the recipient's] deepest interest" on a global scale, fostering

Southern Oregon community members made tiles that combined to form a 120-foot salmon.
Photo courtesy of keithhenty.com.

"personal insight, perspective, and confidence" as well as to develop a more informed sense of international concern.

Also a graduate of the Harvard Kennedy School's "Leadership, Organizing, and Action" program, Sohl is putting her natural talents, her love for the Rogue Valley, and her exceptional training to use. After the exhilarating Salmon Event, with the support of many other Oregonians, she founded Rogue Climate. Rogue Climate's mission, as stated in its website, is to empower southern Oregon communities most impacted by climate change, including low-income, rural, youth, and communities of color, to win climate justice by organizing for clean energy, sustainable jobs, and a healthy environment. The organization works toward its mission through leadership development, political education, and fostering conversations, and it campaigns for policies that benefit communities over the special interests of the largest corporations. Rogue Climate's three primary campaigns are to stop the proposed Jordan Cove LNG export terminal and the Pacific Connector Pipeline (also a strong focus of SOCAN); to transition communities in the Rogue Valley to clean energy; and to pass state-level climate policy that will spur clean energy investment and job creation in rural, low income, and ethnic communities.

It was shortly after Hannah Sohl began Rogue Climate that I got involved in the "No LNG" campaign. I had first heard about Jordan Cove when it was proposed as an import facility back in 2004. Though I thought it was a dumb idea to build a facility in a coastal Oregon earthquake zone to ship fuel to California, I did little other than grumble about it until after the plan changed to export, and our local climate group, 350 Eugene got involved. The group was about to bus to Oregon's state capitol in Salem, and I jumped at the opportunity to join them.

Our big yellow school bus pulled up behind several others just like it, all parked beside the Oregon state capitol. Several dozen of us, all in red shirts, grabbed our banners and posters and clumped down the bus's stairs. Though I had gone to previous lobby days, chatted with my congressperson, and sat in on a legislative session, I'd never demonstrated at the capitol, and I wasn't sure what to expect. But I felt it was important to show up.

I was with Eugene, Oregon's, chapter of 350.org, the organization building global grassroots movements to educate people about climate change, to oppose new fossil fuel projects, to work toward clean renewable energy, and to hold leaders accountable to science and justice. But that particular day the group was lobbying against the most threatening proposal in the state, the Jordan Cove Energy Project.

Much was wrong with this project, but perhaps the biggest issue for 350 was that at the current concentration of carbon dioxide in the atmosphere—well above 400 ppm—and the accelerating speed at which the climate is changing, we simply must not allow any new fossil fuel infrastructure. As the signs and chants insist, we must "keep it in the ground."

We were to gather on the capitol steps at one o'clock, so I had time to wander a bit and see who else was there. I had thought we were all 350 groups from around the state, but as I perused the tables and literature, it became obvious that the Jordan Cove opposition was much broader and more diverse. Several Waterkeepers were there, tribal members, property owners, and a number of southern Oregon groups, which should not have been a surprise. I stopped by the table of a group displaying maps of the proposed Pacific Connector Pipeline, and a signup sheet to "hike the pipe!" In late August and September 2015, concerned folks would hike the route, documenting the ecosystems, communities, and individuals that would be affected by the project. I wanted to join them. Still, though in younger days I had run marathons, now I doubted I could manage a ten-kilometer, say nothing of 229 miles. But I made a note to try to get acquainted with these folks as well as with other activist organizations.

We visited with other demonstrators, admired signs, then gathered on the steps. There we joined in songs and chants and listened to presentations introducing stunning statistics and deep emotion. A landowner on the pipeline access cried out, *I worked my tail off for decades. Put all my money in my land. Some bastard with more money than me wants to take my land away from me. Well, I'm not gonna let 'im do it. He'll have to kill me first. White man took my land once. I'll be damned if I'll let him do it again.*

From the capitol, we walked, signs and banners held high, to the State Land Board where we hoped we might influence the board (composed of the governor, treasurer, and secretary of state) to decide against Jordan Cove's application. I walked behind a huge papier-mâché caricature of Governor Kate Brown and was surrounded by signs with hand-painted blue earths, their messages reading "I Love This Planet" and "Protect Your Home." Other moving messages included "There Is No Planet 'B,'" "Water Is Life," "Save the Bay," "There Is No Second Nature," and "Life Over Profits." A long banner read "Gov Brown, Be a Climate Hero."

I was a little disappointed that no one joined us from within the State Land Board building. But as we spoke and sang and chanted and displayed our signs, I bet someone took the opportunity to peek out a window.

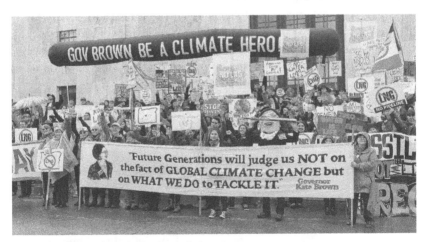

No-LNG Rally in Salem, Oregon's, capitol.
Photo by Allen Hallmark.

And then back home we went.

It's always hard to gauge what difference such a gathering makes. Still, I often remember a little song from grade school days that contained the lines, "Little drops of water/ little grains of sand/ make the mighty ocean/ and the beauteous land." Individual efforts may not have an apparent effect, but many of them together add up to something that can make a difference. Perhaps a whole ocean of difference.

Before the trip to Salem, my impression had been that the opposition to this gas project was localized to the Coos Bay area and a fairly narrow band along the line of the proposed pipeline. I wondered how it had grown so fast and so widely to include people and groups from the whole state. Eventually, I was told that was largely due to Rogue Climate, and the work of organizing whiz Hannah Sohl.

Recent demonstrations at the Dakota Access Pipeline (DAPL) had an effect as well. That 1,172-mile-long pipeline is for shipping oil from the Bakken shale fields in North Dakota to customers abroad. A portion of the pipeline had been rerouted to a corridor abutting the Standing Rock Indian Reservation, from where it would tunnel under Lake Oahe, a major water source for the Standing Rock Sioux. The tribe, upset that the pipeline would threaten their environmental and economic well-being as well as religious, cultural, and historic sites, stood firmly against it. Protests brought Native Americans from all over the nation, becoming the largest gathering of Indigenous people perhaps ever. Countless others, Indigenous or not, and not only from the states but also from abroad, joined the protests.

Though social media had been covering the events, mainstream media was largely absent until bulldozers leveled a sacred site and security workers sicced attack dogs on protestors. Then television as well as YouTube gave the public a short course in pipeline problems and concerns, educating, horrifying, and in some cases empowering the viewers.

Patty and Thomas Joseph of the Hoopa Nation were at Standing Rock, where they organized a kitchen and Hoopa camp, feeding thousands of people over several months. They also attended all of the Jordan Cove informational meetings. The Josephs live on ten-thousand-year ancestral land along the Trinity River, a tributary of the Klamath. "What affects the Klamath affects us," Thomas Joseph says.

The Klamath Tribes, along with Yurok and Karuk, have become intervenors on the Jordan Cove project, giving them legal status to join litigation without requiring permission of the litigants. They and the Hoopa are active in protests and open houses. Tribes wrote an official letter to the Federal Energy Regulatory Commission and another to the State Historic Preservation Association saying that the risks of liquid natural gas to Indigenous cultural and burial sites and to the rivers are unacceptable. Describing the letter to those gathered for a protest, Klamath Tribes chairman Don Gentry stressed that the tribes are concerned about sustainability—about preserving a healthy world for the seventh generation—and are already seeing climate change effects on their plants and fish. And they're seeing increased earthquakes caused by fracking. Gentry urged, "It's now time to stand up and say 'enough.' It's time to stand globally," he said, reflecting the spirit of DAPL. "We need to support . . . other tribes, to work together."

Jordan Cove supporters talked about the jobs the project would create and the tax money that would boost the bottom line for the counties. Hoopa tribal member Patty Joseph countered that most of those jobs would not be available for local people. She reminded those gathered that the fracked gas wouldn't even be for American use; the pipelines would be made from Russian steel, and it was a Canadian company that would profit. Nineteen-year-old Mahlija Florendo from the Hoopa Youth Council added that the Klamath River was not only the home of salmon, but also where her people found food, water, and basket-making supplies. Addressing the people supporting the project, she said, "I just want [you] . . . to know, it's not only going to have an effect on us Indigenous communities, but in every community. Human beings weren't built to live off money."

Cattle rancher Bill Gow lives on the pipeline route. He said he was offered an insultingly low amount of money from the company so they could take twenty-six acres of his ranch, and if he doesn't cooperate—which he doesn't intend to—they will use eminent domain to take it. "How would you like it if I came into your house and took your things?" Gow asked. "Can I do that? Can I walk into your house and take whatever I want? Then why can you do that to me?"

Meetings and outreach inspired many to write to the Federal Energy Regulatory Commission (FERC) with carefully researched and passionate comments on the ramifications of the project and the many reasons it should be denied. More than 25,000 citizens, including ranchers, property owners, youth, businesspeople, anglers, and climate activists as well as the Native American tribespeople, wrote letters critical of the project. The Western Environmental Law Center, which had worked against Jordan Cove for more than a decade, submitted, together with the Sierra Club, comments opposing the project on behalf of a diverse national coalition including commercial fishing, conservation, and private property interests. WELC attorney Susan Jane Brown noted that it was clear "from the sheer length and complexity of the comments" that the project is "legally, ecologically, and socially flawed."

Of course, FERC didn't receive letters solely opposing the project. Colorado's Democratic Governor John Hickenlooper wrote that Jordan Cove would be the only West Coast export terminal that could link Colorado gas to new markets. Gas drilling had been on the upswing in Colorado until an oversupply dropped prices. In the resulting slowdown, many people lost jobs, affecting personal, local, and state economies. A West Coast terminal would let the rigs start drilling again. Utah's Governor Gary Herbert had sent a supporting letter earlier, with similar reasoning. Clearly, a West Coast export facility would be a win for drilling interests and a serious blow to areas such as Colorado's North Fork Valley that are fighting to preserve their land from fracking, as well as being yet another blow to the climate.

Hike the Pipe

Hair on Fire Oregon is a nonprofit headquartered in Ashland, so named to express the urgency of addressing climate change, which is already clearly evident in southern Oregon. In the spring of 2015, Hair on Fire hired Alex Harris, who, with his new degree in environmental science, had returned to his southern Oregon home to find a job fighting the Jordan Cove LNG project. Harris was well versed in climate science and, realizing that Jordan Cove would take climate change in exactly the wrong direction, felt compelled to try to stop it for the sake of his home region and state, for the planet, and for his generation's future. Aligned with Hair on Fire's mission to educate people about the dangers of climate change, he began to develop what he referred to as "backwoods/recreational activism" in rural areas.

Brainstorming with landowners Deb Evans and Ron Schaaf, the three hit on organizing a hike along the proposed route of the pipeline. A hike through high and dry plains, through forests and along rivers; through national forests and miles of private land. For over a decade, concerned Oregonians had written letters, attended meetings, demonstrated, and lobbied their congressional leaders. A hike would be an opportunity to see up close the places impacted, plus to interact with both rural and urban people and communities, and perhaps to help them interact with each other. They hoped it would also encourage the media to share the experience and effects of the Jordan Cove project with the non-participating public. Rogue Climate did the work of obtaining a zillion permits, and "Hike the Pipe," all 229 miles of it, was born.

A couple dozen hardy souls set out on a hot, dry day August 22, 2015, from the rural town of Malin, Oregon, where the proposed pipeline would connect to the existing Ruby line from Opal, Wyoming, and head west. In the week it took the group to cross Klamath Basin's dusty roads, sagebrush grasslands, marshes, and juniper woodlands, lizards scuttled by or did blue-bellied push-ups from nearby rocks, mule deer disappeared behind trees, raptors soared overhead, and occasional waterfowl scouted the route south along the Pacific Flyway. By the end of the week, the hikers plunged joyfully into the cool green forests of the Cascade Range.

On about day ten, the group was joined by dozens more Oregonians to hike together a section of the Pacific Crest Trail, where a highway-size swath would be cut for the pipeline's crossing. The expanded group began one of the major goals of the hike, documenting the old-growth trees that would be cut from near the popular trail to accommodate the pipeline. Hundreds of acres of ancient forest were scheduled to be lost. After this rewarding but also dispiriting documentation of impending catastrophe, the hikers headed for a big rally in Shady Cove, where the pipeline would cut under the Rogue River. They were greeted with cheers, signs, and song and got to float five lovely lazy miles along the Rogue.

Two more rallies would greet the group, one in each impacted county, each rally organized by the community, not imposed from outside. From Rogue Basin's Shady Cove, they would head through the Umpqua drainage to Winston, near Roseburg, and from there, 110 dense and cool miles across the Coast Range to Coos Bay. As the group hiked, they met hundreds of landowners, community members, and activists and listened to some of their stories and speeches.

The line would rip through Bill Gow's 2,500-acre ranch in southern Oregon's Douglas County. "In the 1970s," Gow said, "we were always talking about energy independence. Now they have this gas, and [instead of keeping it for energy independence] they want to export it by stealing my land. It just doesn't make sense." He wanted people to understand that the fight against Jordan Cove wasn't just the work of a bunch of left-leaning tree huggers. "Opposition to this project is across the political spectrum and reaches to the far right."

Neighbors to Gow are Stacey and Craig McLaughlin. They bought their 357 acres in the '90s, at an affordable price because "it was a mess," says Stacey. The land had been logged over and "had four huge dump sites," McLaughlin added. And the house "wasn't really habitable."

The McLaughlins hauled away hundreds of appliances that had been lying about in the woods. Cleaning up trash and delivering recyclables took nearly two years, working almost daily while they also worked full time off the ranch. Then they began restoring the land. They planted thousands of trees, improved damaged wildlife habitat, and put a new roof on the house. It's beginning to look and feel like an undisturbed forest, and they recently welcomed a semi-resident herd of elk. Now some corporation wants to clear-cut a mile across their property, destroying their improvements and restored wildlife habitat.

Stacey McLaughlin says that at first her fight was about protecting their land and the hard work they had invested. But as they learned more about fracking and natural gas, and about climate change, it became much more. She says that as "a small part of this thing we call humanity," they feel a responsibility to the planet and to the future.

Francis Eatherington, a longtime warrior for the forests, is part owner of land along the pipeline route near Medford. In the mid-1970s, Eatherington was one of a group of women who bought 147 acres in those southern Oregon foothills as a sanctuary for women, wildlife, and a biodiverse forest. They named their land Oregon Women's Land Trust, usually shortened to OWL. In the 1980s, as part of an all-women Hoedad crew, Eatherington did the hard work of planting Douglas firs up and down the steep slopes around Tiller, an area also impacted by the pipeline. She has remained a fierce advocate for the forests through the years, working against irrational timber sales with Cascadia Wildlands, Oregon Wild, and other forest conservation groups. As the Federal Energy Regulatory Commission considered permitting the pipeline and other US fossil fuel development, Eatherington joined a mass protest fasting on the steps of FERC headquarters. She was active in a landowner's group that sent informational letters to landowners along the Pacific Connector Pipeline route, letting them know that they had rights and didn't have to sign. Deeply supportive of Hike the Pipe, Francis and other Land Trust women hosted hikers overnight and joined them as they trekked across OWL property.

September 26, five weeks after the intrepid travelers began hiking what turned out to be 310 miles, they arrived in Coos Bay right on schedule. Hikers could hear cheers and music before they could see the hundreds of supporters with signs, song, and food. Hike the Pipe organizer Alex Harris reported that it was "a very powerful image" with the "crowd and their signs spread over the McCullough Bridge and dozens of 'kay-aktivists'" floating below. This moving finale, along with various stops along the route, was well covered by local and national media, perhaps bringing home to a wider populace both global and social costs of a project like Jordan Cove.

The following spring, March of 2016, brought joyous news. FERC denied the project! The commission found the potential impacts to

OWL women celebrate giant madrone, spared from pipeline clearing.
Photo by Francis Eatherington.

communities and landowners along the pipeline route to be far greater than any benefits the project might bring. "When I heard the news I couldn't believe it at first," Stacey McLaughlin said. "I stood at my kitchen sink, looked out the window at our mountain and just started crying, weeping with relief that it was over, finally over. My first thought was I would get my life back, get my seeds planted."

Landowners celebrated. Tribes celebrated. Rogue Climate and SO-CAN, Rogue Riverkeeper, Klamath-Siskiyou Wild, CALNG and so many others were thrilled, gratified, relieved. They hugged and laughed and partied. It had been a long, hard slog, but it had been worth it. Western Environmental Law Center's Susan Jane Brown commented that FERC was usually a rubber-stamp agency, and their denying of this permit showed how seriously flawed the project really was.

But the celebrations were as short-lived as Jordan Cove's second death. Veresen announced that they would appeal the FERC decision, and the 675 landowners began receiving notices of intent to acquire land rights for the pipeline. So the coalition sucked in a huge collective breath and went back to work.

The following December, hallelujah! Jordan Cove was struck down once again. But by then, a new administration had been elected, one far friendlier with the oil and gas industry. The transition team's energy statement said, "We will streamline the permitting process for all energy projects, including the billions of dollars in projects held up by President Obama, and rescind the job-destroying executive actions under his administration." One official specifically promised that the new administration would issue permits to Jordan Cove, which hardly promises unbiased analysis on the part of FERC, the regulating agency whose job it is to evaluate permit applications.

In February of 2017, Veresen announced its intent to reapply, earning the project the zombie moniker. Klamath County landowner Deb Evans said, "We've fought this speculative project twice and won, we'll do it again."

Many members of the opposition held fast to the knowledge that Veresen didn't have the financial reserves or pipeline experience to pull off such a massive project, even after September of 2017, when the company officially filed its new application. Western Environmental Law Center announced landowners, tribes, and community groups' readiness "to stop the pipeline and LNG export terminal for the third time in twelve years." More than four hundred landowners, organizations, tribal members, and concerned citizens filed motions to intervene opposing Jordan Cove, while only five were filed in support, according to a report by Rogue Climate. Then Calgary's Pembina Pipeline Corporation, a fossil fuel giant, merged with Veresen, considerably upping the stakes.

November of the same year, Jordan Cove received another setback when the Oregon Land Use Board of Appeals (LUBA) unanimously rejected Coos County's approval of the energy company's land use permits. Crag Law's Courtney Johnson, arguing for Oregon Shores Conservation Coalition, had pointed out numerous errors in Coos County's analysis. In its decision, LUBA noted that the county had not weighed the project's impacts on commercial and recreational fishing and access to shellfish beds against perceived public benefit of the project. LUBA also faulted the county for ignoring adverse impacts to shore areas from construction and methods for mitigating damage to wetlands. Responding to statements from the Confederated Tribes of Coos, Lower Umpqua, and Siuslaw, LUBA stressed that input from local tribes must be considered during the land use review process.

Oregon Shores Conservation Coalition is well into its fifth decade of helping citizens preserve the coast and its ecosystems while protecting the public's access. The group works through grassroots education and citizen science as well as through formal land-use processes, such as LUBA's, and occasionally through the courts. Since 1971 when it began, Oregon Shores has grown to over a thousand members who watchdog and protect everything from the crest of the Coast Range to the edge of the continental shelf, looking at watersheds, intertidal areas, beaches, and headlands. More recently, the coalition also includes a strong focus on studying and preparing for sea rise and other climate-induced changes.

Climate Impacts

Early in 2018, Oil Change International published the first publicly available comprehensive analysis of Jordan Cove's climate impact. As it compresses fracked gas to liquid, the terminal alone would emit 1.8 million metric tons (MMT) of greenhouse gas annually, becoming the worst polluter in the state after 2020 when Boardman, Oregon's, last remaining coal-fired plant is scheduled to be shuttered. The project's Oregon emissions—the facilities in Coos Bay plus the pipeline—would be about 2.14 MMT. Tracking the lifetime greenhouse gas emissions, including typical leaks during drilling, transporting, liquefying, shipping, and storage plus the gas's eventual burning, the group reported Jordan Cove Energy Project would be responsible for the emission of 36.8 million metric tons of greenhouse gas per year. This would be 15.4 times the amount from Boardman, an amount equal to putting an additional 7.9 million passenger vehicles on the road, or nearly two more cars for each Oregonian.

With legislative action in 2007, Oregon had announced its intention to reduce climate pollution to 10 percent below 1990 level by 2020 and at least 75 percent below 1990 level by 2050. And after President Trump backed out of the Paris Accord, Oregon's Governor Kate Brown accepted the challenge for Oregon to uphold Paris. The more stringent goal set in Paris—that global temperature rise should be kept to 1.5 degrees Celsius—requires global fossil fuels at zero by 2050; for a maximum 2 degree rise, emissions need to hit zero by about 2065. Jordan Cove's emissions would render impossible Oregon's stated climate goals. And it would contribute an ever-increasing percentage of the state's proposed emission allowance, without even providing Oregon with any energy.

SOCAN co-facilitator Alan Journet was particularly pleased to see the Oil Change International report because it reflected and confirmed calculations that he had sent to FERC in 2014, in comment on their inadequate draft environmental impact statement. While ostensibly evaluating environmental impacts, FERC ignored lifetime effects, cumulative effects, and monetary calculation of social cost—the costs caused by the project to our economy, health, and the environment.

When considering climate change, it is disingenuous not to consider the full lifetime of greenhouse gas emissions, Journet points out. Whether GHG is emitted at drilling, in leaks on route, at the terminal in Coos Bay, or when ultimately burned in Asia, it is equally culpable in thickening the atmospheric blanket. FERC has likewise ignored the cumulative effects of installing a new pipeline. Once the construction expense has been incurred, the company would obviously want to keep maximum supplies flowing, requiring more drilling, encouraging projects such as those being fought in Colorado's North Fork Valley. Jordan Cove would lock in decades of increased gas and oil exploration, drilling, transporting, and burning at a time when fossil fuel use must be phased out.

Dr. Journet also stresses the importance of considering calculable social costs. In August of 2016, a federal court officially ruled legal the calculation of long-term economic damage, estimated through an interagency process at $40 (but ranging from $12 to $116) incurred from each ton of carbon dioxide or its equivalent released into the atmosphere. Using the low end of the department of energy figures, costs would range from about $400 million to well over $600 million per year. The high-end figures bring annual costs to more than $6 billion. Whatever limited economic benefit might derive to a local tax base cannot compare.

But the bottom line is that the earth is already well on the way to the upper limit of global temperature increase. National and planetary climate goals require the power sector to be de-carbonized by 2050, meaning that we should now be phasing out all fracking and new gas infrastructure, not bringing on new projects.

Opposition groups have been frustrated that state politicians as well as Oregon's national delegation have either supported Jordan Cove for perceived economic reasons or have been silent on the issue. But at about the time of the Oil Change International report, US senator Jeff Merkley, a vocal climate champion, took a politically brave stance and publicly opposed the energy project. Early in this century when Jordan Cove was proposed as an import station, many people still considered natural gas to be a bridge fuel between coal and renewables. But you can't fight a fossil fuel problem with fossil fuel. As Sierra Club's executive director Mike Brune said a few years ago of natural gas, "It's not a bridge; it's a gangplank."

It is true that natural gas is much cleaner *burning* than coal. For at least a couple of decades however, any unburned methane that escapes into the atmosphere is more than eighty times as potent as carbon dioxide at capturing heat, and the drilling and transporting processes are notoriously leaky. Currently, 25 percent of human-made global warming is from escaped methane. With little knowledge of methane's effects or of the leaks, toxins, and pollution prevalent in fracking, it is understandable that political leaders could imagine Jordan Cove as an economic engine as opposed to an environmental disaster. But now the facts are much better known.

Senator Merkley originally saw the project as a possible answer for an area suffering from losses in sawmill jobs. He still wants to build infrastructure and help the southern Oregon coastal economy, but not at the expense of landowner rights and the climate. Jobs and the environment cannot be an either/or dichotomy. People and their liveli-hoods must never be ignored, but a healthy environment is essential for people, as well as for ecosystems. The goal is to find good jobs while fostering a healthy earth.

In an opinion piece in Medford's *Mail Tribune* (12/7/2017), Merk-ley notes that Jordan Cove would become Oregon's biggest carbon pol-luter and says, "With a renewable energy future within reach, it makes little sense to help Asia leap from coal to natural gas, locking in carbon pollution for decades."

Merkley adds, "That carbon pollution is a big deal because it is driv-ing up global temperatures, with a new record virtually every month. We are feeling the impact of climate disruption across Oregon. It has killed a billion baby oysters; disrupted our salmon fisheries; and gener-ated worst-ever droughts, hurting farmers and ranchers," along with contributing to catastrophic wildfires through nearly a half million acres in 2017 alone.

Though Senator Merkley's opinion piece had numerous detractors on economic grounds, Jordan Cove opponents cheered his statement and hoped Governor Brown and other officials would follow suit. It is a tenuous hope though, as focused as many are on short-term eco-nomic figures above all others. Early in 2018, the Portland Business Alliance (PBA), representing 1,900 member businesses, wrote a let-ter to FERC supporting the project. *Street Roots News* reporter Rick Rappaport commented in a March 2018 article that he shouldn't be surprised about PBA's support because "I kind of figured that those in charge of the PBA would support Attila the Hun if he promised jobs burying the dead," but he was shocked that businesses professing to be sustainable and environmentally friendly would go along with such a position.

And that seems to be where the problem lies. Big business and corporations have the money to fight citizen opinions and initiatives. And many citizens feel their lack of buying power more than they feel or fear the changing climate. Probably a majority of people understand that climate change is a fact, but few understand either the ramifications or the urgency—the fact that once tipping points are reached, which could happen in the next very few years, there is little if anything we can do about it.

So that is where the Riverkeepers and Hair on Fire, Rogue Climate and SOCAN come in. They are fighting Jordan Cove because it, like other new fossil fuel projects, takes us in precisely the wrong direction. But their greater goal is to educate the public to the fact that urgency exists, and that each of us can help if we act now and act together. SO-CAN's initials are for Southern Oregon Climate Action Now. A pie chart on their website shows all the areas in which they work. Around the edge of the pie, it says, SO CAN YOU, SO CAN WE, SO CAN ALL OF US.

The "all of us" piece is the key. A common complaint is that groups sharing the same goals often don't coordinate and may not even be aware of one another. They duplicate efforts, wasting human energy, or sometimes actually work at cross-purposes, competing for support or recognition. SOCAN is part of a coalition of climate-oriented groups. Those in southern Oregon meet quarterly, sharing information on projects, goals, and concerns. And they are part of regional, state, and national climate groups. Rogue Climate and 350 help environmental groups throughout the state and across the nation connect with each other. Riverkeepers and the Waterkeeper Alliance also benefit from keeping in touch with each other regionally and internationally. More and more, groups are recognizing the importance of weaving a broad net. All political persuasions, religions, ethnicities, economic levels are together in this ever-warming ever-deepening bathtub. With cooperation and coordination, we can fashion the best possible float—and perhaps even turn down the temperature a little.

SOCAN has made a documentary featuring various southern Oregonians telling how climate change affects them personally. The last interviewed was Dan Wahpepah, Anishinaabe and Kickapoo founder of Red Earth Descendants, a coalition of regional earth-based cultures. He talks of problems in the rivers, forests, and struggling species, then tells of an Iroquois prophecy foretelling a time when the trees will be dying from the tops down, the bad will get worse, and the good will get better. Encouraging membership in the latter group, he said, "Earth is our church," explaining that water and biodiversity govern life and are truly holy. People have lost their way, he adds, and must come into their spirituality by listening to their elders, their brothers and sisters, the plants and animals, the birds and trees.

Those are foreign concepts for much of twenty-first-century money- and technology-based America. But in more rural areas, maybe especially those such as southern Oregon with its forests, mountains, wild rivers, and proximity to the ocean, larger numbers of people are still connected to the natural world by occupation or inclination. Perhaps that is where it will first be recognized that the earth is holy. The best jobs and fattest wallets won't do much good without it. And that is where Jody McCaffree, Robyn Janssen, Hannah Sohl, Deb Evans, Stacey McLaughlin, Francis Eatherington, and so many others will continue to fight for a livable earth, now and in the future.

Getting It Together, Together 5

The Beginning

IN AUGUST OF 2004, TROPICAL BIOLOGIST Jason Bradford moved to Willits, California, a rural town at the southern edge of California's redwood forests. Exploring Willits Environmental Center, nearby Hopland's solar-energy-boosting SolFest, and Ecology Action's workshop on biointensive farming, he found many kindred souls. Here were people concerned about both the general acceptance and the precariousness of the consumer culture, and looking for more viable ways to live.

We've gotten ourselves in quite a predicament, they all agreed. Besides everything else that's going sideways, most of what we require for survival comes from far away. Our food is shipped thousands of miles; we're importing water as well, since it took water to grow the food. We import building materials; we import energy—in wires, pipelines, and tanker trucks. So what happens when disasters strike, natural or manmade? What happens when transmission lines are down or highways buckle? Plus, those energy sources we're using are finite. At some point, they will become inaccessible or gone. What then?

The answer seemed clear. People must stop relying on getting everything they need from elsewhere. And they must evaluate how much they really need, as opposed to just want, to slow the depletion of natural resources. They must instead work toward making their local society sustainable.

Jason Bradford arranged community showings of the documentary *The End of Suburbia*, which depicts the suburbs as a dream of idyllic life that is dependent on the car to actualize. Issued in 2004, the year after a major East Coast energy blackout, the film distrusts the reliability of infrastructure-tied energy, inspiring many to question ways of life they had formerly taken for granted. With increasing numbers turning out for each showing—thirty, then sixty, then ninety—Bradford connected with numerous people eager to search together for effective ways to secure the kind of lives they and the community wanted, honoring people and the planet, and, to the extent possible, providing locally for their needs.

Basic needs for humans are the same as for other animals: food, water, breathable air, shelter. Because humans are a social species, they also require community. For warmth, light, cooking, transportation, and operating machines, they look for a source of energy; to market goods and services, they need a form of currency for exchange.

The citizens of Willits had experienced the potentials of advocacy and action during years of fighting Whitman Corp's Remco Hydraulics' dumping of chromium and other toxins used in their chrome-plating operation from the 1960s through the '90s. Seemingly random dumping on the ground and in nearby waterways in this flood-prone area had contaminated soil, air, and groundwater and was the presumed cause of numerous local health problems. The citizens eventually won a settlement for cleanup as well as damage awards for hundreds of people's medical issues.

So once again, they banded together, a dozen or so people forming Willits Economic LocaLization (WELL) Project. The community took inventory of local talent, then formed study groups and began action toward localizing energy, food, water, and transportation. Participation grew, and the model spread to other communities.

One of the people Jason Bradford met during his showings of *The End of Suburbia* was Brian Weller, a self-described resident alien, come to Willits from Great Britain with a background in facilitation, group dynamics, business development, and applied cognitive science. In 2006, two years after WELL's launch, Weller traveled to Port Townsend, Washington. Somewhat isolated at the northeast tip of the Olympic Peninsula, on a thumbnail of land thrust into the Puget Sound, Port Townsend is home to many long active in the search for positive change. Judy Alexander, Deborah Stinson, Steve Hamm, and others had been zealous members of the peace movement since 2002, before the United States invaded Iraq. Frustrated and anxious as they watched the George W. Bush administration exchange the nation's post-9/11 compassionate unity for a polarizing drive to war, the group felt compelled to work double time for peace. But by around 2005, Alexander's focus on the peace movement became conflicted as she began to see worldwide wars, ostensibly fought for democracy or human rights, as actually only thinly disguised pursuits for power or resources. If acquisition and exploitation were the driving forces of war, those impulses had to be addressed and turned around for any hope of a harmonious world. Alexander understood that unless Americans learned how to live with a healthy biosphere as their organizing principle, simply working for peace would be futile.

Rick Van Auken, Judy's brother, along with Hamm, Joe Breskin, and others, formed the Jefferson Energy Center, a group pushing to get local public control of energy, at that point provided by the Puget Sound Energy (PSE) Company. Meeting at their regular "energy lunches," the Citizens for Public Power approached the local water and sewer People's Utility District (PUD) about taking back the energy services from PSE. They put the idea to a vote by Jefferson County citizens and in 2010, two years after the successful passage of the proposition, the local PUD bought the county's electrical system from Puget Sound Energy and arranged for power from Bonneville Power. Encouraged by community response to localizing energy, when the Energy Center heard of WELL's work they were intrigued and requested more information.

As the Energy Center's guest, WELL's Brian Weller explained how in working together, the Willits community was finding local ways to ensure resilience in times of disasters such as earthquakes or oil shortages. In so doing, they were gaining confidence earned through self-reliance and the emotional support of working with others of like minds. His message of local self-sufficiency hit exactly the right notes in Port Townsend.

Alexander, Hamm, Deborah Stinson, Bill Wise, Linda Kay Smith, and the energy group joined forces with other interested citizens and, inspired by the WELL Project, designed their own venture that they named Local 20/20 for its connotation of clear-eyed vision along with a goal of attaining local resilience by the year 2020. Their mission became to work together toward local sustainability, integrating ecology, community, and economy through action and education. Energized citizens selected study groups according to their interests, including emergency preparedness, food, transportation, climate, energy, waste management, and economic localization. And work began.

Meanwhile, in 2004 in another port city, far away in Kinsale, County Cork, Ireland, permaculture designer and teacher Rob Hopkins was wondering how to maintain viable civic life without the help of fossil fuel, which currently powers most everything. Looking for diverse insights, he challenged his students to consider the question: how can we apply permaculture principles to the design of community living?

Permaculture design is informed by the function of natural ecosystems. It begins with living soil, nourished by water, the water's flow protected by plants plus the health and the contours of the land. Entire life cycles of plants, animals, and animalcules—their reproduction, growth, death, and decomposition—are honored and used. There is no waste stream, no outside input except what drifts in naturally in wind or water or traveling animals. Diversity within an interconnected complex system adds greatly to resilience.

An impressive example of this design approach is the Chikukwa Project in Zimbabwe, the largest permaculture site in the world. In the early 1990s, photos from that area showed only an occasional tree, barren hillsides rutted with erosion gullies, soil around springs compacted by cattle hooves, and homes lower on the hillside half-buried by silt that heavy rains had washed from the hill above. Soon the main springs dried up and villagers had to walk five or more kilometers downhill to find water. The populace was sick, hungry, and discouraged.

Patience Sithole, now the administrator of the project, remembers the deterioration and despair following the failure of the spring. She sat in the kitchen with Eli Westermann, a German woman who came to the village to teach, along with Eli's teacher husband, Ulli. "We sit down, seeing our land going," Sithole said. "If it rains, we see . . . soil going. We discuss. We always discuss. And then one day said OK, but we have to [have] action. And we began action."

About ten local volunteers contoured the hillsides, planting vetiver, a hot-climate grass, on berms at terrace edges. They gathered seeds and planted native woodland around gullies and springs as well as on the ridges of most hills and some slopes. The terraces and ridges, plants and their roots ensure water filtration during the wet season, and its gradual release during dry months. The trees also provide firewood and timber.

Villagers planted their own gardens and orchards. They climbed hills, digging swales and cleaning out springs. Ulli Westermann recalls, "It was very enthusiastic; it was a beautiful time. 'Cause we were working with the community, singing and digging away."

The site responded quickly, and as villagers saw what their neighbors had done, they did their best to replicate it on their own properties. It was perhaps an advantage that Zimbabwe's economy collapsed in 2005, giving no impetus for villagers to head to the city for jobs. If a living was to be made, it had to happen at home.

Monthly *Permachikoro* (permaculture school) classes spread to all of the six villages in the area. In just a few years, rather than barren or muddy hills, the site was filled with small farm households, each surrounded by orchards and vegetable gardens, and spring water that bubbles up year around for each of the villages.

Crop yields are way up, nutrition is diversified, hunger and malnutrition are no longer the norm, and the locals feel invested and self-reliant. Fertilizing the soil with animal manure and compost, farmers needn't worry about the expense of modern chemical fertilizers and pesticides. Protecting water sources and using indigenous plants makes plantings relatively self-maintaining.

With knowledgeable local people using appropriate plants and gardening methods, Chikukwa didn't see the fallout typical of projects when outsiders come with textbook solutions and then return home. Too often in poverty abatement programs, foreign methods and materials are imposed upon a culture with the idea that modern (and western) is better. In contrast, a basic permaculture principle is that sustainable agriculture must be adapted to the specific climate and culture. In Chikukwa, appropriate plants were used, and the natural methods resonated with the local climate and culture, where elders had been raised to revere the soil, plants, and water. Whereas projects prescribed from the outside rarely last more than a few years, the Chikukwa Project still flourishes after more than two decades.

Julius Piti, a Chikukwa Project founder, says, "Permaculture actually solves all the problems that we face in human life. So this is . . . the right approach for us to live—if you want to save the earth."

Common problems that we face in human life, and certainly in community projects, are chaotic emotions: hurt feelings, blame, defensiveness, misunderstanding. Phineas Chikoshama, from the Chikukwa BCCR (Building Constructive Community Relations), recognizes that conflicts are everywhere. Individuals have conflicts within themselves, he says, and if they've no peace within, how can they avoid conflicts with others? The BCCR recognizes three "circles of knowledge" for community relations: indigenous knowledge (how would our elders handle this?), spiritual knowledge (what answers do your beliefs provide?), and analytical knowledge (what does your education tell you?). A community discussion ensues, combining the three circles, and a resolution follows, avoiding shaming. To blame and shame points fingers but doesn't solve problems. Understanding a transgressor, rather than accusing and blaming, allows him back into the community. Everyone can grasp the cause of the problem, and together they can find a solution.

When sociologist Dr. Terry Leahy of the University of Newcastle, New South Wales, and his documentary-making wife, Professor Gillian Leahy of University of Technology, Sydney, visited the Chikukwa to hear and film their story, they were able to sit in on a two-day conflict resolution session. An erosion gulley had formed, apparently because a villager had overharvested trees for firewood. Water was flushing soil from some yards and depositing it in others. Villagers who had been involved in the original berm making and planting resented the damage. The villager who had cut the trees felt he was just gathering firewood, as he had always done. Some of the people remembered that at their *Permachikoro* classes they were being taught how to handle such problems, and requested a Conflict Workshop.

The community event began with prayers and singing. Then members of the BCCR staged a humorous skit of the conflict, explaining motivations in a nonjudgmental way. Amid much laughter and discussion, villagers drew their understanding of the conflict. The next day, the group decided that they would repair the gulley with rocks, contouring and planting more vetiver grass; they would prohibit more cutting in the area and plant more trees. Then they all adjourned to the mountain and reconstructed the terrace together. Having learned how to accept the potential for conflict and provide a dispassionate and inclusive response undoubtedly contributes to the long-standing success of the Chikukwa permaculture project.

Guided by permaculture principles usually focused on a landscape, Rob Hopkins's students in Kinsale, Ireland, set out to design a functioning, self-regulating, zero-waste, localized ecosystem for people. They considered multiple aspects of a community: energy production, health, education, economy, and agriculture. Two students, Louise Rooney and Catherine Dunne, looked at the current situation in each area (e.g., that most of Ireland's food was imported, noting the energy required to grow and ship) and described a vision of how to satisfy that need locally. In the case of food, the need could be satisfied locally through establishing orchards and community gardens in parks and public greens, developing a food co-op and farmer's market, encouraging residential edible landscaping, and supporting new agricultural systems. Rooney and Dunne developed a detailed plan with columns for what is, what we'd prefer, and how we accomplish it. Their plan would become the Transition Towns concept, which the Kinsale Town Council decided to adopt as it worked toward energy independence. The word *transition*, of course, describes the process of changing from one condition to another. It would take a major transition—a shift both in technology and attitude—to segue from lives dependent on fossil fuels and other outside resources to lives more independent and interdependent.

Hopkins moved back to his hometown of Totnes, England, and there he and Naresh Giangrande developed his students' concept into the Transition model. In 2006, Transition Town Totnes was born, the first Transition Initiative. The term "Transition Initiative" came about as the concept was adopted and adapted in communities ranging in type and size: villages, neighborhoods, boroughs, streets—not just Transition Towns. Within a year, Hopkins began the Transition Network, which trains and supports people and disseminates concepts of Transition Towns. As of June 2019, 1,200 Transition Initiatives were active in fifty countries.

Meeting Marissa

As these movements were beginning in Willits, Port Townsend, and Totnes, and as the Chikukwa gardens and gardeners in Zimbabwe were maturing in health and resilience, a sensitive and impassioned teenager headed to the University of Wisconsin-Madison to find her life's purpose. Tuned in to suffering, both in the world at large and in her own backyard, Marissa Mommaerts resolved to fight her gnawing depression by finding a way to be of service to others. At Madison, she studied international relations, which brought her to Nigeria to learn the Yoruba language.

"After visiting Nigeria," Mommaerts says, "I could never look at the world the same way. The perverse imbalance of power between the US and the countries from which we extract resources to fuel our consumer culture was so blatantly clear. I saw people starving, dead bodies on the side of the road, miles and miles of slums and homeless people. A woman asked me to take her baby and bring him to America."

While pursuing her master's degree in International Public Affairs, Mommaerts began a sustainable development project in Lake Victoria, Uganda, but soon realized that foreigners going to someone else's community to fix their problems reeks of what she calls the "White Savior Complex," plus they just aren't effective. So she decided to try working at the international level, hoping she might be able to help increase justice and opportunity for the so-called developing countries, bringing them some degree of symmetry with that of western industrialized nations.

She landed her "dream job" at a Washington, DC, think tank where she grappled with worldwide challenges such as human rights abuses, poverty, population growth, climate change, resource scarcity, and war. She worked with world leaders, visiting four continents in six months. Her parents often didn't know which continent she was on, Marissa laughed, "but my mom would give up eating meat [Marissa is a vegetarian] whenever I was abroad, as a sacrifice to the Travel Gods to protect her most radical offspring."

Though her job was stimulating and rewarding, she soon realized that, "as long as this pervasive global power imbalance perpetuates widespread exploitation and injustice for the benefit of a gluttonous few, as long as we desecrate our exquisite natural environment, as long as I have brothers and sisters around the world for whom suffering is the norm rather than the exception, I cannot sit at my comfy desk and wait for marginal, insufficient progress from the international political system I have exalted for the last seven years."

So Mommaerts left her job and searched again for her life's path. She "unplugged" (put away her electronic gadgets), slowed down, opened her eyes, took a deep breath. She gardened, cooked, repaired, and "began to view the world around me with gratitude and wonder, discovering a sense of peace that comes with living simply, connected to a community and the Earth."

It was in this receptive state of mind that Mommaerts discovered Transition. Here at last were people who could clearly see the plight for society and the biosphere triggered by following the current economic model, but rather than merely resisting or railing, they were rolling up their sleeves and building a better way. Working in community, they were creating generous, joyful, effective lives less dependent on fossil fuels or other, often polluting, outside resources.

Mommaerts would become the new program director of Transition US (TUS), the national hub for the International Transition Network. TUS seeks to inspire and support an ever-growing national network of local resilience powered by community action. "Resilience," in this context, is the ability of a community to regain its former vigor and efficiency after any major assault, such as floodwaters, earthquakes, or other stress to the community or its infrastructure. To quote from the Transition US website, "[B]y unleashing the collective genius of our communities it is possible to design new ways of living that are more nourishing, fulfilling, ecologically sustainable, and socially just."

As Mommaerts traveled from Washington, DC, to the TUS office in Sebastopol, California, she visited Transition leaders through the Midwest and South, admiring seed libraries that assured maintenance of proven and adaptable lines in spite of the increase in seed patenting by powerful companies; communities developing green jobs and energy-efficient, affordable housing to combat urban decay and gentrification; groups learning how to finance community-owned renewable energy systems. She was thrilled and uplifted by their work, and was welcomed by them as if she were family or a longtime friend.

Arriving in Sebastopol, she moved into the cozy TUS offices, tucked above a ballet studio and a mattress store, and took to heart the symbolic contrast from her upscale Washington, DC, office, with its view of DuPont Circle.

Mommaerts charged into her new life as not only TUS program director, but also as a member of the Transition Sebastopol community. She joined with others to hang a clothesline, plant three gardens, build a new compost facility for a community health center, and organize a neighborhood food and plant swap. Savoring all of that plus work parties every free night and weekend, Marissa felt overjoyed to be part of a community building local resilience.

Then in 2013, at an event that was part of Transition Town originator Rob Hopkins's US speaking tour on "The Power of Just Doing Stuff," her eyes locked on the azure eyes of a fiery redhead named Jeremiah. For the next several months, those two kept crossing paths—at Transition Sebastopol meetings, again at the several-day Village-Building Convergence. Once, when Mommaerts was facilitating a Transition Sebastopol meeting with twenty overly lively participants, Jeremiah helped the group refocus. "That caught my attention," Marissa says. She remembers thinking, "He's not only cute, he can handle group dynamics!"

Six months after their eyes first met, Jeremiah brought Marissa to one of his favorite places for their first date: an old-growth redwood grove near the California coast. "It wasn't until then that I had witnessed his passion for nature and his knowledge of plants," Marissa said. That was more than a little attractive as she thought about gardens built and gardens dreamed of. *I love a man with a green thumb*, she thought.

One of Marissa and Jeremiah's early dates was to hear Winona LaDuke speak about Indigenous resistance to invasive fossil fuel infrastructure, such as fracking wells and pipelines. The native tribes were putting their safety on the line to fight against something endangering not only their immediate environment with the loss of habitat and the potential of spills and pollution, but also the entire planet, as burning of more fossil fuel exacerbates climate change. Moved by LaDuke's challenge to the audience to support or create anew such activism, Marissa and Jeremiah were drawn closer as each saw their own fervor echoed in the other.

A Visit to Port Townsend

At about this same time, Port Townsend's Local 20/20 joined the International Transition Network, connecting them with TUS. When I talked with Marissa, she suggested Local 20/20 as a Transition group I might like to visit to see how specific needs were localized.

David and I had driven the three hundred miles to Bremerton, Washington, from our home southwest of Eugene, Oregon, to visit his brother, as well as high school friends of both his and mine en route. It was just a quick hop farther to drive up Kitsap Peninsula, bridge across Hood Canal and up the Olympic Peninsula to Port Townsend, tucked above an expansive bay. A charming town of ten thousand, its buildings clustered on a hill looking through sailboat masts to a seemingly endless sweep of water, Port Townsend watches the Strait of Juan de Fuca flush into Puget Sound with each day's high tide.

In the nineteenth century, this seaport town had expected to become a major West Coast city and built elegant Victorian buildings appropriate for that future. But when a depression hit, stopping the railroad track construction at the south end of the Sound, that projected destiny was foiled. Stately homes and public buildings emptied as their occupants sought greater opportunities. Eventually, the railroad line continued, but up the east, not the west side of the water, through Olympia, Tacoma, and Seattle, rather than to Port Townsend. With

Port Townsend, Washington, waterfront.
Photo by Jason S. Squire.

fewer people and no push for development, the Port Townsend Victorian buildings remained, adding this historic architectural style to the natural charm of the seaport.

Local 20/20

Local 20/20's mission statement begins with "working together," its original and ongoing focus. The group partners with other nonprofits and civic groups furthering the mission of increasing local self-reliance and sustainability. Spreading their umbrella increases local resilience by meshing redundancies and expanding resources, energies, and connections. Projects in multiple areas wax and wane organically.

Like an ecosystem or permaculture garden, all aspects of life need to be considered and woven together into a strong and resilient system. L20/20 action groups include, but are not limited to, localizing food, energy, economy, emergency preparedness, climate action, transportation, and waste.

Localizing Food and Farming

Richard Dandridge, Local 20/20's transportation chair, had set up a meeting for David and me at Judy Alexander's comfortable home. An early focus for Alexander, basic for any person or group trying to localize their lives, centered on food systems. *How shall we feed ourselves?* The food on a chain-store shelf travels an average 1,500 miles to the store, requiring burning fossil fuels for that trip on top of the fuels used in industrial agribusiness. In order to travel well, produce must be bred to be tough and have long shelf life rather than being tasty and nutritious, but fragile. Earthquakes, floods, or other emergencies can collapse highways, stopping or delaying food delivery. And supermarkets stock only a three- or four-day supply.

Industrial agriculture requires uniformity of crop and growing methods, reducing biodiversity, depleting soil, polluting waterways with chemicals, and requiring considerable fossil fuel use in fertilizers and power equipment. Most meat and produce brought in from outside the area is raised industrially. Approximately 80 percent of US antibiotic use is for farm animals raised in crowded and unsanitary conditions. Residue remains in the meat ingested, exacerbating the antibiotic resistance that is increasingly common in the health system. Port Townsend's 20/20 wanted clean meat, reliable nutritious produce, healthy soil and waterways, and a thriving agricultural economy, all of which would seem to mandate local production.

Their Local Food Action Group began by promoting a network of neighborhood-based community gardens, which by June of 2009 numbered more than twenty. These gardens included rows or whole gardens grown specifically for local food banks as well as ample space for home use. Sharing with folks unable to garden seemed basic. Most gardens were planted and harvested collectively rather than having each gardener be responsible for her own plot, giving sociability, support, and mentoring to what otherwise could be an isolating, unfamiliar, or tedious task.

With citizen participation high, the group's focus shifted to getting acquainted with local farmers and their products and learning how the community might best support these farms. Inspired by a sustainability class Alexander had taken through the Northwest Earth Institute (NWEI) in 2009, her local NWEI group launched multiple simultaneous discussions on local food, the discussions stimulated by film showings. Working through the farmer's market and grange, it set up more than twenty groups throughout the area, each with a local food producer (farmer, fisher, cheesemaker) who could describe the work and challenges of producing food and depending on local support for income. At the end of the discussion series, a nearby farm hosted a potluck where group members shared their epiphanies and intended behavior changes. Through the townspeople's new understanding, Port Townsend saw a huge, immediate, and gratifying turnaround as customers and income ballooned for the farmer's market, Community-Supported Agriculture (CSAs—subscription to regular food boxes from local farmers), and the food co-op, a trend that continues and has spread. Local farms, food businesses, and food security thrive in Port Townsend.

Food is so bottom-line essential for any functioning community that I looked for other examples of how food systems could be fully maximized. On the TUS website, I found an inspiring example of local work supporting food justice. A group in Florida has seen terrific success with a project that helps consumers, producers, and the climate as well as the volunteers. Transition Sarasota's Suncoast Gleaning Project envisions a route to a world without hunger. Since 2010, it has contributed over a quarter of a million pounds of organic fruits and vegetables to the local food insecure—those lacking sufficient resources for household food, of whom there are forty-two million in the United States. Each Monday morning, volunteers convene at an organic farm to gather whatever the farmer has been unable to sell that week and deliver it to the All Faiths Food Bank—save for a shopping bag full each volunteer may take home. It's a quadruple-win program: the food banks get fresh organic food; the volunteers get a week's supply of produce as well as the satisfaction of their physical contribution; the farmers get a tax write-off; and the atmosphere is saved from the addition of methane that would have been released from wasted food rotting in a landfill.

Only 27 percent of Sarasota County residents live within a half-mile of retail healthful food. The USDA estimates that twenty-three million Americans, including six and a half million children, live without access to quality nutritious food. Inspired by the success of the Suncoast Gleaning Project, the Florida Department of Agriculture is preparing a gleaning pilot project in six counties. The food bank has expanded its distribution manyfold.

Organizer Don Hall says that the project is highly replicable, but requires partnerships with multiple organizations and people. He recommends building relationships with food banks, farmers, and soup kitchens, and educating the public on the importance of local food. To that end, the Suncoast Gleaning Project sponsors an Eat Local week, with farm tours, farm-to-table dinners, film-screening, and keynote speakers.

꿀

And gardens can provide opportunities beyond the gustatorial. A lush and expansive garden lies behind Churchill High School in Eugene, Oregon, nearly tropical-looking in its vigor. Bees, birds, and butterflies enliven the vibrant scene. Marissa Zarate, executive director of *Huerto de la Familia* (the family garden), leads a group of us, explaining the history and objectives of the program, which was launched to alleviate hunger and social isolation among local Latino residents.

Huerto offers hardworking and ambitious Latinos who have been systematically disadvantaged—through workplace discrimination, lack of education, extreme poverty—a place to grow their food, plus training in organic gardening, food preservation, and nutrition. At the same time, the garden provides a setting for families to work together, where parents can pass on cultural heritage and self-sufficiency skills to their children, as well as an opportunity to connect with the larger community. Organic garden program manager Danielle Hummel says her passion is "building community through food and farming."

Begun in 1999 by a small group of volunteers with one garden plot to serve six women, *Huerto* by 2018 had expanded to six large plots spread throughout the Eugene-Springfield area, serving eighty-five families.

We wander among tomatillos, hot peppers, cilantro, and corn, enjoying the fragrances and appreciating that people are able to grow culturally appropriate crops, save hundreds of dollars, and improve their health as they eat *well*, not merely what they can afford. Marissa Zarate tells us about the six-week *Siembra La Sena* (Seed to Supper) class, giving *Huerto* students every aspect of growing, harvesting, preparing, and preserving these good vegetables. But the opportunities don't stop at the garden fence or the dining table.

Huerto also offers training in small business creation and management through their *Cambios* (change) Micro-development Program, providing life-changing training in business skills and resources and in financial literacy. Most of the local Latinx population is employed in seasonal or low-paying jobs, so the shift from uncertain employee to business owner is a dramatic one, psychologically as well as financially. Some years, those studying food production businesses can get hands-on experience at the *Huerto* food booth. After training in restaurant business operation, tracking finances, display and marketing, and health and safety regulations, they design menus, prepare food in a certified kitchen, and sell to the public from the booth. Other years, *Huerto* focuses on another phase of food marketing, the fundamentals of operating a produce stand for a small farm business. In 2018, a *Huerto* farm stand showcased irresistible produce at Oregon's Lane County Farmers Market.

Cities, too, can act as catalyst between farms, farming, and consumers. In Belo Horizonte, Brazil, where 38 percent of the population was at or below the poverty line, farmers' bottom lines improved along with public health when the city committed to supporting the doctrine of food as a right. The city encouraged community gardens and farmer's markets throughout the urban area, established limited price controls, making the local produce more affordable, and subsidized certain cafés where citizens could enjoy an affordable meal from local produce. A stunning resulting statistic from the stable increase in availability of good local food was a *72 percent* reduction in mortality of children under age 5, from 1993 to 2005.

And the growing of that food can have rewards beyond farmers' income and the nutrition of the food. Many speak of psychological benefits—the stress reduction and pure pleasures of working in the garden, certainly a lifelong experience of my own. One particularly moving story I heard was from a Vietnam vet who said that he cured his PTSD by gardening, attributing his recovery to his intimate connection with the garden's ecosystem as he worked barefoot and gloveless in his rich organic soil, teeming with life-giving microorganisms. He felt the sacredness of the soil and for the first time, felt himself to be truly connected to the rest of the natural world—perhaps the first time he had felt he belonged anywhere.

That vet's cure seemed miraculous to some, but studies have shown soil bacteria, specifically *Mycobacterium vaccae*, to uplift attitudes even of people having incurable diseases such as tuberculosis and, yes, PTSD, at least partly through acting against stress-induced brain inflammation. See studies by Christopher Lowry, Mathew Frank, Mary O'Brien, and others. The healthful effects can be gained by working barefoot in good organic soil, ingesting the microbes on food or in supplements, or by breathing them in on a long healthful walk in the woods.

Love Also Grows

Marissa Mommaerts and Jeremiah Garcia celebrated their first Valentine's Day, which was almost the first anniversary of their connection, with their hands in the dirt "in service to the soil and water and plants and animals that give us life, and in service to all future generations—including, we hope, our own children—who will inhabit this planet someday."

The couple spent the morning planting willows to help restore a floodplain along Maacama Creek, a tributary of the Russian River. Working together, they held a common vision of the benefits of their work: A healthy floodplain lessens concerns of urban flooding damage, which National Oceanic and Atmospheric Administration estimates to cost $8.1 million per year; it improves water quality, thus improving the health of humans as well as of other animals; and it supports natural vegetation and macro-invertebrate communities that are essential to the lives of birds, fish, and other wildlife.

In the afternoon, they worked with the contours of their own backyard garden in true permaculture fashion, helping water infiltrate and building fertile soil to grow plants for food, medicine, and habitat tough enough to thrive, as Marissa says, "regardless of whether future tenants have green thumbs."

As the day drifted to a close, Marissa marveled at what a deeply fulfilling day it had been, "pouring sweat and love, patience and tenderness into . . . these islands of a future we know is possible." A silent prayer of dedication formed within her: "May our love be a vessel for healing ourselves and our world."

A healthy floodplain contributes to the health and development of stream-flow dynamics and of riparian plants and animals. So also do organic gardening methods and produce contribute to the health and development of the soil community, and the health of both the atmosphere and the human community. And as the stage was set for the health of the ecosystem, the relationship of two human beings doing the regenerative work grew in health and vigor as well.

Localizing Transportation

When the car that had delivered Marissa to Sebastopol broke down, its days were numbered. She vowed to use fossil fuels as little as possible, enjoying getting stronger, biking wherever she could, supplementing with public transit, and borrowing a car if absolutely necessary. "It's really liberating," she says. "It feels so good!"

Local 20/20's Richard Dandridge signs his emails, "Pedaling On!" A committed cyclist, he shared newspaper clips with me from Local 20/20's files, some having to do with cycling, some with other ways of getting around without the use of fossil fuels. The Port Townsend and Jefferson County newspaper, *The Leader*, runs monthly columns for Local 20/20, a great service for any group trying to extend its influence. Several recent columns were about 20/20's Active Transportation Study Group. Port Townsend's Lys Burden wrote about the local monthlong celebration of active transportation including riding the bus, as well as walking and cycling. The month's events boasted a bike tour of the city, a Bike Fest, celebration of National Bike to School Day, contests earning school awards, Bike to Work Week, and free instruction and use of tools at the ReCyclery each Friday and Saturday throughout the month of May.

Burden came naturally by her interest in Port Townsend's active transportation month. She and her husband, Dan, along with Greg and June Siple, were featured in a 1973 *National Geographic* issue bicycling from Alaska to Guatemala. On their 18,000-mile hemisphere-long trek, the group planned a nationwide, cross-country "bike centennial" for 1976 to commemorate the bicentennial of America's Declaration of Independence. Over four thousand cyclists turned out, cycling all or portions of the 4,250-mile event.

Group cycling is becoming an increasingly popular business event as well, replacing the boardroom or golf course as a place for colleagues to bond or potential clients or partners to negotiate. Beyond the returns of health and pleasure, leading, drafting, sharing efforts, anticipating problems on a ride can give insight on or help train similar behaviors in the business world.

But pedal power isn't the only non-fossil-fuel way to get around. Thomas Engel, retired University of Washington chemistry professor and moderator of Local 20/20 Energy Action Group, wrote a column in May of 2017 about electric vehicles. Most people know that a gasoline-powered car belches out tons of climate-damaging CO_2, but how many know that they can save as much as a thousand dollars a year in fuel costs by driving an electric vehicle instead of gasoline-powered? Or that if you charge your car in town with PUD power, your money stays local, while if you buy gas at the pump, your money decamps to distant countries and big corporations? Or that most modes of transport have now been electrified? Buses commonly are electric; some eighteen-wheelers have hydrogen fuel cells that make electricity to drive the wheels, one prototype a class 8, with up to 2,000 horsepower and the ability to haul 80,000 pounds. The biggest hang-up with fuel cells is lack of infrastructure, but the Nikola company has plans for a network of 364 fueling stations across the United States and Canada. Also, aircraft hybrid power is being developed, and Norway has electric ferries powered by lithium ion energy storage systems.

Engel noted that Tesla, famous for luxury electric vehicles (EVs), by early 2016 had reservations for 300,000 of its relatively affordable Model 3 EV. Battery charges are getting longer, with a Chevy Bolt good for 238 miles on a single charge while until recently EV single-charge range was less than 100 miles. And EVs have been on the market long enough that it is now possible to buy a three- to five-year-old Nissan Leaf with a range of 60 to 80 miles for about a quarter of the original purchase price.

In a "Think Resilience" online course, author Richard Heinberg, senior fellow at the Post Carbon Institute, suggests that the best way to design a resilient community is to focus on transportation. If the layout is comfortable and safe for walking or biking, if necessary shops or medical facilities are available in accessible nodes, if mass transit is convenient, comfortable, and accessible, that goes far toward community resilience. And if you add the engaging component of contests and festivals, as Port Townsend's Local 20/20 has done, or encourage businesses to capitalize on the advantages of group cycling, the participating numbers explode.

Localizing Finances

Marissa Mommaerts is typical of many who wanted to live sustainably but found it difficult within the current system, particularly in a high cost-of-living area such as Sebastopol, California. In Marissa's case, student debt, a nonprofit's wages, and high living costs tested her resources. Having multiple roommates helped. So did living simply, growing, preparing, and preserving food and other do-it-yourself projects. Work-trade can be a boon for many, where an established person with a roomy house can offer rent-free space to an industrious, often young, person willing to help out. This of course—like most everything—works only when agreements are clear and both sides possess goodwill and communication skills.

Mommaerts refers to Vicki Robin's update of her book *Your Money or Your Life* (2018), which guides readers to evaluate, step by step, their relation to money and its effect on the quality of their lives. Robin follows this by listing ways to spend less so that you don't have to work so hard, and ways to find and pursue the things that bring your life true pleasure and satisfaction rather than mere money and discontent. It's basic economics.

The word *economy* evokes thoughts of money, but its origin is the Greek *oikos* (house) and *nemein* (manage). So economics has to do with managing one's home. Mommaerts and other leaders of the Transition movement, along with an ever-increasing number outside of Transition, see the current globalized, consumer-driven economy (called a "banks and tanks economy" by climate-justice organization Movement Generation) not just as irresponsible home management, but more critically, as the root of our current social and ecological crisis.

One of the most potent ways to improve the management of our home, Mommaerts notes, is to transform our local economies from a focus on profit for a few to one that truly serves people and communities. And it is essential to remember that "ecology" is the study of that same home or household. Once we understand how the household works, we can emphasize what it needs and get rid of what gums up the mechanism. Ecosystems, which we cannot live without, require respect and regeneration, truly an essential "bottom line."

In 2014, Mommaerts co-authored (with Ken White and Ben Roberts) "Weaving the Community Resilience and New Economy Movement Report," pointing out that "there are many alternatives to the mainstream economy: cooperatives, local currencies, barter, sharing, gift economy, participatory budgeting, reclaiming the commons, and so much more." A strategy for local economic transformation she promoted through Transition US has been dubbed "REconomy" (for REsilient and REthink), an approach originated in Transition Town Totnes, UK. The criteria for REconomy enterprises are to produce local goods for local people, to minimize waste and pollution, to respect resource limits and maximize renewables, to provide decent livelihoods and affordable and sustainable products, to treat and pay workers fairly, to practice democratic governance and build common wealth, and to strengthen resilience in the community as well as in the business itself.

As TUS Program Director, in 2016 Mommaerts issued a "REconomy Project Report" of twenty-five enterprises exemplifying REconomy principles. Economic resilience is a critical component to local self-sufficiency. People need to be protected from the vagaries of an economic system subject to inflations and crashes, and they need to have a measure of autonomy, rather than being dependent or beholden.

Investing

An ingenious Local 20/20 project for localizing investment was featured in the REconomy Project Report. American investments, lumping savings in stocks, bonds, mutual and life insurance funds, total about $30 trillion. Not even 1 percent of those investments assists local small businesses, even though half of the nation's jobs and private economy production comes from those same small businesses. But each dollar spent in the community circulates locally within area businesses, multiplying its effect in the local economy. With that concept in mind, and inspired by a visit from Michael Shuman, author of the book, *Local Dollars, Local Sense: How to Shift Your Money from Wall Street to Main Street and Achieve Real Prosperity* (2012), Port Townsend's Local 20/20 became intrigued with the concept of local investing.

At a meeting in 2006, a group of Port Townsend citizens discussed excluding a big chain store that was threatening to come to town. But former Mayor Michelle Sandoval said she didn't want to continue reacting to problems. She'd rather be proactive and support businesses the townspeople wanted, rather than fighting those they didn't.

The dual push to create prosperous local businesses and to keep investment money in the community gave birth to the first-of-its-kind Local Investing Opportunities Network (LION). Between 2006 and mid-2017, LION members (local people having capital to invest) had infused about $7 million into Jefferson County businesses and nonprofits. Neither a loan company nor investment fund, LION creates opportunities for local businesses, nonprofits, and members to network and determine individually if they can work together. Though there are risks, and some of the businesses have failed, investors take heart in the numbers of local businesses that are thriving. Investor Earll Murman says that as he drives through town, he counts fifteen LION-assisted businesses on one particular corridor. Together, those fifteen businesses have 140 employees profiting from LION investment.

Jordan Eades, co-owner of Hope Roofing, says Hope experienced a 50 percent per year growth, serving 1,879 customers in the four-year period of 2012 to 2016, since its first loan through members of LION. During those years, the company paid $5.2 million in wages to employees supporting twenty-six kids. Many other local businesses tell similar stories.

Recently, Michelle Sandoval was approached by some successful young people who wanted to invest, but not in Wall Street. They were concerned about the effect their money would have on the world. LION was the perfect answer, helping local business, keeping the money working in the community, and giving the investor a wonderful, almost parental feeling of helping to launch something worthwhile. LION has become a model for other communities across the nation and internationally. Its key documents for local investing are available online at www.L2020.org/LION.

Alleviating Debt

Across the water in Seattle, a group of Gen Xers turned their attention to the financial burdens of college graduates. In the decade between 2003 and 2013, student debt increased more than 300 percent. Over 70 percent of college students take out loans, with average indebtedness at graduation more than $29,000. Such debts can affect graduates' work and living choices and delay their independence. They may be unable to pursue entrepreneurship or to save for the future. Even when motivated to do so, they may feel unable to work in nonprofits. Negative effects expand from the graduate to the community and the nation. As tuition rises and wages continue to stagnate, talented students from middle- and lower-income families are priced out of higher education.

In 2014, Salish Sea Cooperative Finance (SSCoFi), also featured in the REconomy report, formed to refinance State of Washington students' high-interest loans by reinvesting its members' financial resources and returning the profits to its members. Member-owners of SSCoFi are Washington residents who "wish to participate in reclaiming the credit commons." They include borrowers (graduates who already have student loans), partners (people who want to support the mission), and investors looking for a socially responsible way to invest money in their community, as they lighten the burden of student debt on young community members.

SSCoFi's stated vision is "a world where all graduates will have the freedom to pursue their goals, contribute to the community, reach their full potential, devote their skills to worthwhile causes, and impact the world free of the crushing burden of excessive student debt. [They] are connecting successful, socially conscious individuals with the change-makers of the future." The cooperative enhances its support to debtor-members through fiscal education, mentorship, and community connection.

SALISH SEA (Southern)
Washington and British Columbia

0   50   100 mi

The Salish Sea is an intricate network of waterways separating the southeast end of Vancouver Island, British Columbia, from the northwest corner of Washington State. It continues south with Hood Canal separating Washington's Olympic Peninsula from Kitsap Peninsula and includes other Salish Sea waterways. Port Townsend is at the peninsula's northeast tip. Seattle is south, on Puget Sound's east edge.
Map by Imus Geographics.

Local Currency

Eight hundred miles farther down the West Coast, Oakland, California-based Bay Bucks (also featured in the REconomy Project Report) is a worker-owned cooperative with business members being stakeholders. Railing against the fact that the national monetary system furthers economic disparity at the same time that its growth imperative drives environmental destruction—depletion of resources, over-fishing the ocean, mining mountains, and spewing greenhouse gas emissions into the atmosphere—San Francisco's Chong Kee Tan and others explored the benefits of local currencies. In 2012, they founded Bay Bucks with Transition SF to "create a functioning regional currency that helps local businesses thrive while promoting collaboration and building community wealth." Chong Kee Tan says Bay Bucks' goals are to democratize wealth, protect the commons (those elements and systems we all require such as air, water, and land), and create humane and need-based economic relationships rather than feeding corporate profits.

As a complementary community currency spendable only at member businesses, 100 percent of Bay Bucks stays within the community, stimulating local business and teaching consumers how their money circulates. Because it is zero-interest, Bay Bucks circulate faster, resulting in greater local economic vitality, but without triggering the growth imperative. Therefore, it eases the exploitation of natural resources. Because Bay Bucks are spendable only locally, local labor, materials, and consultation are sought out, further stimulating the local economy. And Bay Bucks gives a hedge against economic recessions. With a local currency, business can continue even if banks are faltering.

A terrific boon to beginning entrepreneurs, Bay Bucks can provide an immediate market within its membership, and it can help alleviate costs as well. Cash flow is typically problematic for start-up businesses. Nomadic Ground organic, fair-trade coffee owner Thomas Landry says that through a Bay Bucks member he was able to get professional consultation to grow his business. Adam Parks used Bay Bucks for graphic and web design services for his organic meat shop, saving limited cash for supplies or other business needs. Parks said having Bay Bucks members as customers was invaluable for starting his business. It delights Holly Minch that she is able to be a customer of her clients—businesses that her Lightbox Collaborative counsels.

Members also benefit from Bay Bucks' commercial barter system, which encourages them to offer their excess inventory into exchange, and withdraw other businesses' excess goods and services as needed. The exchange reduces or eliminates economic loss, provides an economical route to acquire needed goods, and, as it avoids waste, decreases resource exploitation. Members say that local currency can encourage shopping with local businesses, improve a business's bottom line, give communities more say in the direction of local economies, and foster human relationships.

On the Move

Marissa Mommaerts and Jeremiah Garcia also sought ways to protect the commons and create humane rather than exploitive relationships. They dreamed of "a world with drinkable water and breathable air, a world where food is nourishing and our food system isn't controlled by profiteers, a world without debt slavery, extreme violence, or pervasive depression," a world where their children would have a good quality of life. They wanted, as nearly as possible, to bring that world into fruition themselves. As much as they loved the Sebastopol area where they met, and loved their community of dear and like-minded friends, they could not afford farming land in a walkable community in Sonoma County, California.

Instead, they decided to head for Wisconsin and plant a food forest on a small piece of marginal land Marissa's parents own. They wanted to grow as much of their food as possible and gradually extract themselves from the "exploitive global consumer economy." So they converted a 6 by 12-foot cargo trailer into a tiny house on wheels that they would pull behind their new crowd-funded work truck and park beside their new garden, saving money on rent. Once they had planted their Wisconsin community garden, the plan was to join Indigenous activists resisting oil and gas drilling and shipping in Canada, and then travel around the country to work on regenerative agriculture projects, healing the food system, rebuilding topsoil, and sequestering carbon. Their goal was to plant at least a few hundred thousand trees before having a child some unknown years in the future.

They arrived in Wisconsin on May 30, 2016, just in time for Marissa's mother's birthday. And just in time to discover that Marissa was pregnant.

Like many of the best-laid plans of mice and men, their strategy was derailed. Marissa worried that motherhood might make it impossible to be an activist. Her father met the news with, "Too bad the world's such a mess. "

"It's a scary time to become parents," Marissa agreed. She and Jeremiah knew well the climate threats, the societal implications as people compete for scarce resources, and the current and potential ramifications on other countries of the western growth obsession. "But," she said, "the gift of living in uncertain times is that it makes the present that much more precious. The gift of living simply is that it creates more space for the things that really matter. And the gift of having compassion and empathy for feeling the pain and suffering of the world as deeply as I do is that this capacity is matched by a capacity for love."

By the time the garden was finished, the couple had devised a new plan. Access to local organic food in northeast Wisconsin would be difficult, so they would head for an off-grid straw bale home in a sunny Colorado valley renowned for its organic farms, the Valley of the North Fork of the Gunnison River on the west slope of the Rocky Mountains. They headed south and west to once again build community, plus a very special nest of their own in Paonia, Colorado, where Marissa would soon find herself on the Board of Citizens for a Healthy Community, fighting fracking. On her website, there is a beautiful picture of Marissa and Jeremiah standing on each side of a summer-gold path in their adopted high-altitude semiarid home. Their shared gaze warms the viewer as Jeremiah looks across the path at Marissa, her hands cradling the blue-painted globe that is her burgeoning belly.

As for her fear that motherhood would interfere with activism, Marissa says, "that's not possible. Because in order to do my job as a

Marissa and Jeremiah's babe-to-be means the world to them.
Photo courtesy of Marissa Mommaerts.

mother, to ensure my baby has the best chance possible of surviving and thriving in this world, I'm going to have to work harder than ever before. . . . I'm talking about ensuring he has access to the very basics required for human survival: clean water to drink, clean air to breathe, and healthy food to eat. In a world rife with hate and callousness and immense suffering and the potential for so much more suffering, the only thing that is true—the only purpose for existence—is love. And you can bet we will teach this to our son."

Localizing Preparedness

In the sprawling green H. J. Carroll Park south of Port Townsend, Washington, live music plays amidst food vendors and twenty-five booths offering information on disaster resilience resources. In a central area, people are carrying water and shucking corn, two of the activities in an "earthquake relay," a simulation of collective community response to a disaster.

This is the annual All-County Picnic, co-sponsored by Port Townsend's Local 20/20 and the Jefferson County Department of Emergency Management to foster organized neighborhoods, community resilience, and education for preparedness.

Most Transition Initiatives as well as other groups focused on community health and security emphasize the importance of teaching people to imagine what they might need if they suddenly should be on their own. In any large disaster—witness recent fires, floods, tornadoes, and earthquakes—government services are quickly overwhelmed. Particularly in isolated communities, but anywhere, really, community members are far better off if they've planned ahead how best to survive and to come to each other's aid. Local 20/20's Danielle Turissini tells about the importance of preparation at the individual and family level, which, as in the airplane where you're to put on your own oxygen mask first, leaves you in a better position to then help others. If disaster should strike, she says, "It will be up to us, you and me, to take care of each other. . . . It's imperative that we prepare our minds for what to expect and do, our pantries for what we'll need, and focus our hearts on how we can help others do the same."

The Neighborhood Preparedness (NPREP) Action Group working with Jefferson County Department of Emergency Management has trained more than a hundred self-organized neighborhoods since 2006 in classes by Heather Tanaka. NPREP also partners with the

Port Townsend Food Co-op for an annual storewide special pricing and education event, featuring food and supplies necessary for disaster preparedness. Local 20/20 provides a free online preparedness course at www.getemergencyprepared.com. People can check the TUS site for emergency prep information. The number-one thing, which is addressed in all of the action groups, is personal resilience. If you are self-reliant before disaster strikes, you'll be far more likely to get through a disaster if one should come.

Local 20/20's Climate Action group is studying the potential effect of a rising sea by watching the behavior of "king tides"—particularly *high* high tides that happen predictably several times a year, but most dramatically in the winter. Marine meteorology instructor Dave Wilkinson tells of a late winter 2016 king tide that was amplified by wind and a low-pressure system. Predicted to be at about nine feet, it actually topped eleven, leaving seawater and debris in downtown streets and parking lots. Local monitoring stations show that for the past several decades, sea rise has occurred at a rate of about six-tenths of a foot per one hundred years. Future sea rise, however, is projected at two or more feet by 2100. At that rate, a king tide coinciding with a storm would likely "render streets impassable, damage shoreside buildings, and require substantial cleanup," Wilkinson says. The king tides are giving early warning, and the locals are paying attention.

All of NPREP's goals converge at the annual All-County Picnic with talks, booths, and literature about emergency preparedness. While getting informed, citizens can enjoy music, dance, and good food. Picnics, festivals, food, and music are essential ingredients for all aspects of community building. Social connection is a basic value of Transition Initiatives. Respect, inform, include—and celebrate often.

Localizing Waste Reduction

If it can't be reduced, reused, repaired, rebuilt, refurbished, resold, recycled, or composted, it should be restricted, redesigned, or removed from production.—Pete Seeger

Seeger's quote tops Local 20/20's Beyond Waste action group webpage. A basic permaculture principle is that there is no away. Everything is somewhere and a healthy system makes use of all parts of the life cycle. In a four-page document called "Moving It On," Local 20/20 lists "local resources for donating, recycling, and selling your stuff." It includes information on recycling sites, thrift stores, and Internet sites, and provides a detailed alphabetical list from appliances and art supplies to vehicles and yard debris. Beyond Waste is currently researching waste recycling and remanufacturing initiatives and technologies related to food and pet waste, composting and soil fertility. It also has a number of new links on ways to live without plastics, an urgent current concern.

By the early 2000s, the world output of plastic waste rose more in a single decade than it had in the previous forty years. Eight million tons of plastic end up in the oceans annually. A plastic garbage patch, estimated by the Ocean Cleanup Project to cover 1.6 million square kilometers, lies in the north central Pacific Ocean. Such plastic can remain in the environment for centuries, breaking down into smaller and smaller pieces, often ingested by fish or other animals that mistake the tiny plastic pieces for food. Those animals can then die from lack of nutrition or the plastic-filled creatures can become part of our own digestive systems. Ninety percent of seabirds contain plastic debris and many feed it to their young. Other animals get tangled in nets or six-pack rings. Plastic is also now routinely found in salt, and in most of the world's tap water. Reusing containers and recycling or disposing properly rather than tossing would be a good place to begin to address the plastic-waste problem. Finding a way to do without should be a next step.

As Seeger says, everything should complete its life cycle so that we have no waste. A critical modern problem is wasted food. Forty percent of US food is never eaten according to a 2012 report issued by the Natural Resources Defense Council—that's sixty metric tons a year, with an estimated value of $165 billion—while one in eight Americans struggles each day to have enough to eat. American families throw out about a quarter of the food and beverages they buy. More waste comes from farms, cafeterias, stores, restaurants—all areas of the food chain. The UN's Food and Agriculture Organization says that food discarded by retailers and consumers in the most developed countries would be more than enough to feed the world's 870 million hungry.

Beyond the immorality behind those facts, the environmental effect of food waste is considerable. Food production consumes a tremendous amount of water, fertilizer, and land along with the fuel to process, re-frigerate, and transport the produce. In the United States, 40 percent of those resources and costs are therefore wasted, along with the waste of the food itself. Add to that the fact that a large percentage of that wasted food ends up in the landfill where it emits methane, a greenhouse gas many times more potent than carbon dioxide. Someone quipped that if landfills were a country, they'd be third only to the United States and China in their production of greenhouse gases. Unlike many problems, this one could be, if not solved, considerably improved, by distributing good unused food to the needy, teaching consumers how to avoid waste, and wholeheartedly promoting composting. Compost is excellent garden fertilizer and, unlike inorganic commercial fertilizers, isn't made with petroleum products. Compost completes a plant's life cycle and is a local product.

I saw an exemplary community solution at a thriving neighborhood garden in Eugene. The garden was tended by a dozen regulars and that many more irregulars, its produce shared generously, feeding both the heart and the stomach. But what delighted me the most was the community compost. Two large bins stood in the green right-of-way strip at the end of the block. In one, neighbors deposited waste from their yards or their kitchens, covering kitchen waste with leaves, sawdust, or grass clippings stored in the second bin. As the material decomposed, it was used on the neighborhood garden, and no neighbor on that street needed to send food or other organic material to the landfill.

As in many urban environments, there can be problems with rats being attracted to food, but Judy Alexander says that needn't stop community composting. Mesh bottoms, with smaller than half-inch grid for containers—plus mesh lids or even a central mesh cylinder inside a regular bin, the space filled with insulating sawdust—keep away most hungry rodents. Keeping the compost moist discourages critters looking for a warm dry nest. Or use a Bokashi bucket in the kitchen, fermenting scraps for a couple of weeks before adding them to the compost. Rodents steer clear of Bokashi ferments. For single-family use, trench composting may be the simplest method. Dig a hole in an unused, well-drained garden spot and toss in the scraps, then cover. No odor. No rats. And the garden will thank you.

Up the Sound from Port Townsend, a municipal recycling company in Bellingham, Washington, solves the problem nicely for anyone not wanting to collect compost for their own use. FoodPlus! is an award-winning program available for residential and commercial customers alike. It collects food scraps, garden debris, compostable paper, and other organic materials, delivering them to Green Earth Technology in nearby Lynden, to be made into compost available for purchase throughout the county or beyond. The FoodPlus! program reports that 25 percent of material in the landfill could be composted. Green Earth composts 20,000 tons of material a year, saving local businesses and residents dump fees, saving landfill space and methane emissions, and providing excellent organic matter for gardens and farms.

Local Skill Sharing

Pointing toward the self-sufficiency required for resilience, Transition groups look for a revival of prowess once common but now fading from familiarity. Bellingham's Transition Whatcom hosted its first annual SkillShare Fair in September of 2012. Nearly six hundred locals turned out to learn forgotten work techniques while surrounded by music, Irish dancers, barbeque, and curly fries. In booths and hourly demonstrations, people taught skills such as the use of a scythe, how to make a camp stove, how to keep bees, raise or milk goats, do basic carpentry or gardening, how to start a business or darn socks. Eighty-nine-year-old Erma Boothby said, "I never knew so many people would be interested in learning how to darn a sock! Even men!"

A mini-aerobic digester was a popular demonstration. Aerobic digestion is a bacterial process that reduces volume and changes composition of organic waste such as food, horticultural products, or sewage, giving it the potential for a new life as excellent organic matter in the garden. Talented people willing to share and a public eager to learn, reminisce, or just enjoy each other's company coalesced to make the skill fair a big success.

Now called Cascadia SkillShare and Barter Faire, the Bellingham event continues the teaching and learning of lost skills, showcasing some new ones, and providing a place to meet and learn from community folks with experience in these crafts and trades. The "barter" part of the Faire is to allow people to trade their surplus items (handcrafted, wild-harvested, homegrown, tools, building materials, toys, clothing) for things they need.

Repair Café is a closely aligned movement that also harkens back before the dawn of our throwaway culture. Born in the Netherlands in 2009, it now has spread to more than 1,300 locales in thirty-some countries.

Decades ago, if a zipper broke or a part wore out people fixed it or had it fixed rather than tossing it out for a new model. The perceived need of constant upgrade is a major driver of resource depletion. Gathering repair people in a festive atmosphere, Pasadena, California, holds monthly Cafés, now with fifty repair people and needing twelve volunteers to organize and help manage each event. They focus on a particular kind of repair each month, with townspeople swarming in, their arms full. There's a month to bring the collars to be turned, the seams to be sewn, the jacket to be altered. Another for the iPad or other electronics, a month for appliances, another for bikes, or for furniture or most anything you can imagine.

From Pasadena, Cafés spread to Vermont, Maryland, Texas, New York, and beyond. Similar events go by different names, such as Eugene's Fix-It-Fair. Some Cafés add services like knife-sharpening, seed-swaps, haircuts, live music, trade markets, and, like Skill Shares, resilience-building classes such as composting, worm culture, and chicken raising. At the end of the day, attendees are happy and have saved money; repair people have made money and connections as well as passed on expertise; the community feels more connected; and the landfill fills more slowly.

Skill share and repair Cafés are particularly delightful concepts to me. Being a depression-era baby, brought up by parents raised on farms and a mother who had early learned how to "make do," making dresses from flour sacks, darning socks on burned-out lightbulbs, working a buttonhole, home-pressing apples for cider, patching worn sheets, and so much more were everyday facts of my growing up. I feel privileged to have learned basic self-sufficiency as a child and am gratified now to see respect for such activities returning. It is a great comfort to know you can do it yourself if it is an important piece of day-to-day life.

Building Community

The next time David and I are exploring around the Olympic Peninsula, I want to continue north a little farther. The Lopez Community Land Trust, another enterprise listed in the REconomy Project Report, is on Lopez Island, Washington, the third largest of the San Juan Islands in the Strait of Juan de Fuca, not terribly far from Port Townsend.

Sandy Bishop and Rhea Miller started the land trust with about a dozen others in 1989, responding to a shocking 189 percent increase in housing prices on the island. Many individuals and families were being priced out of housing, and when people lose the right to have a home, they tend to lose access to community as well. Bishop and Miller founded the Trust to provide "permanently affordable, environmentally sensitive and socially responsible community development."

The Trust built highly praised developments in the 1990s in response to the need for affordable housing. When they began contemplating their fourth project in late 2005, Executive Director Sandy Bishop recalls that she was mulling over a talk given by William McDonough, an architect, designer, and author of *Cradle to Cradle* (2002), a book showing how in imitating natural cycles, people could create a no-waste society.

McDonough asked, "How do we love all of the children of all species for all time?" as he explained the "Hannover Principles," statements that he and Dr. Michael Braungart formulated to guide sustainable design. The first principle reads, "Insist on the rights of humanity and nature to coexist in a healthy, supportive, diverse, and sustainable condition."

"He says *insist*, not hope," Bishop pointed out, as LCLT sought a stronger focus on the environmental component.

She and Miller initiated a three-day design charrette, inviting designers and practitioners from all disciplines of housing and land development. Out of that brainstorming, they selected Mithun Architects, a firm that states, "We believe in design's vital capacity to connect people to place and each other." Mithun seemed ideal to carry forward McDonough's Hannover Principles and the Lopez Community Land Trust goals.

How It Works

Homeowners purchase shares of the Trust land with sweat equity as well as with cash, and build with the support of more than one hundred construction interns and local skilled tradespeople. For some of the first-time homeowners, building is a novel experience. Bishop mentioned her pleasure seeing a woman who had never before even held a measuring tape, now busy on her roof with a nail gun. As homeowners help build their straw-bale homes and learn new skills, they not only become more emotionally invested in their homes, they gain empowering feelings of self-reliance.

Building together with other community members and university interns, homeowners also build trust of former strangers from far different backgrounds. Many of the new homeowners had never lived with the support of a community. Participating in work and decisions develops what are increasingly endangered skills that are necessary for taking part in a democracy—skills like listening, advocating, and arbitrating.

The Lopez Community development features green building (building that minimizes impact on human and environmental health), water catchment, solar hot water, grid-tied solar power, passive solar heating, and a goal of zero net energy. Zero net energy necessitates an energy-conservation mindset on the part of the householder, which takes education and monitoring and for many, adaptation, but most householders are eager to learn, both for the ideal and to save money.

Regenerative Agriculture

Rhea Miller and Sandy Bishop met in 1984 on a five-thousand-mile peace walk across the nation. Five thousand miles give lots of opportunity to wonder, observe, evaluate, and get acquainted. In a June 2018 blog, Miller remembers their shock, as they were walking through Missouri, to discover that in the previous year in that state alone, eighty-nine farmers had killed themselves. As they walked by boarded-up rural businesses and empty farmhouses, Miller compared her rural Iowa childhood in the 1950s and '60s, a time when children explored meadows and woodlands with no concerns about toxic dumps or trespassing charges and visited warm and welcoming family farms that raised pigs, chickens, sheep, and cattle plus a whole rainbow of diverse crops.

Miller, whose childhood was before the days of corporate agri-biz, quotes the Center for Disease Control that in seventeen states tenant farmers and other agricultural workers currently have a suicide rate five times greater than the general population. Such shocking statistics along with the need for good food, healthy soil, affordable land, and knowledgeable farmers able to provide good quality of life, worked together to inspire the Lopez Community's commitment to sustainable agriculture.

LCLT's Sustainable Agriculture and Rural Development (SARD) program is committed to regenerative agriculture and land stewardship. The program works with local farmers and volunteers to support small-scale ecologically sound land-based livelihoods. It teaches about the importance of resilient local food systems, promotes the use of local food in the schools, and provides school-based food and gardening education. LIFE (Lopez Island Farm Education), begun by SARD but now part of the Lopez Island School District program, connects students K–12 to the natural world and teaches nutrition, land stewardship, community, and the environment.

Elementary students garden weekly, learning each step from seed planting to harvest with teachers who connect the work to literacy, social studies, math, and science. Middle-schoolers grow produce for the school cafeteria, using three tunnel greenhouses as well as the outdoor garden, and by now may be harvesting from the new orchard. By high school, students can elect a class at the nearby fifty-acre biodynamic (an agricultural method treating soil fertility, plant growth, and livestock care as interrelated and important both ecologically and spiritually) Stonecrest Farm, where they not only prepare soil, plant, transplant, and harvest, but also tend farm animals and learn to make butter, cheese, yogurt, vinegar, and sausage. Another popular elective is a culinary class in the school kitchen. The SARD program recently purchased Stonecrest Farm, assuring the ability to continue and expand its work in farming opportunities and education.

Begun as a response to the discriminating effects of exorbitant home and real estate prices, the Lopez Community Land Trust remains committed to community and opportunity for home ownership, by 2018 beginning its seventh project. Once the Trust idea germinated, it has grown and blossomed into a place that develops the human economy, health, and spirit as well as the human's essential connection to the greater community of the biosphere. The Trust honors its commitment to build a vibrant world for "all of the children of all species for all time."

Reclaiming the Commons

"Relationships are at the heart of Transition," Marissa Mommaerts says, "and for people to get to know each other, they need places to gather. Sadly, many communities and neighborhoods lack ample public gathering spaces." Addressing that need, Transition US sponsored a webinar on the Village Building Movement by Portland, Oregon's, "City Repair."

Not to be confused with Repair Cafés, which repair *things*, City Repair aims to mend *communities*, helping people connect to each other and to their ecological place through community artistic, horticultural, or natural building projects. Its stated mission is to foster "thriving, inclusive, and sustainable communities through the creative reclamation of public space"—or reclaiming "the commons," defined as "land or resources belonging to or affecting the whole of a community." Active for over two decades, City Repair's inspiration came both from Native American wisdom and twentieth-century political events in Portland.

In the 1970s, Portland's Planning Commission instigated a design competition for a public square in downtown Portland, next to the Pioneer Courthouse. The mayor, however, opposed the plan, preferring an enclosed gathering space that would require admission fees. Recalling that freedom of assembly requires a place to assemble, and aware that there were no public plazas in the city, planners and the winner of the design competition for the proposed square decided to press the issue. With broad public backing, a cross-cultural collaboration successfully claimed the square as a public place in defiance of the will of the city administration. More than ten thousand people turned out for the dedication of the Pioneer Courthouse Square in April of 1984.

Citizens responded enthusiastically to the exercise of public power, and the young people who would become City Repair's founders watched the change of spirit engendered by the claiming of public space. Now City Repair helps communities coalesce through projects like painting a street intersection as a plaza, or with permaculture or natural building projects. As a community decides together what it wants and works together to make it happen, personal confidence, new relationships, and community cohesiveness develop, "awakening every person to their own capacity and destiny," as a Lacandon Maya elder foresaw of City Repair's work.

It was at a 2014 City Repair Village Building Convergence in Sebastopol that Marissa and Jeremiah got better acquainted. Transition US worked with Transition Sebastopol and the new Sebastopol Village Builders for the ten-day convergence. Since then, the Village Builders have continued to inspire and support place-making projects, and City Repair continues to spread its work.

In June of 2018, the organization held its eighteenth annual Village Building Convergence in Portland, a ten-day event presenting workshops, panels, and performances on community building. It featured more than thirty place-making projects throughout the city in intersection repair, ecological landscaping, and building with cob, a natural building medium made from subsoil, water, and straw or other fibrous material. In these workshops, everyone is invited to participate, and if special skills are needed, as in cob building or permaculture gardening, teachers are available. Imagine the spirit and energy engendered by a ten-day festival learning about and participating in the creation of beautiful places—stunning plazas, benches, gardens—that can be visited, shown off. The experience remembered: *I helped make that. We built that. Together.* And together the people reclaimed places where they could gather, get to know each other, and become functioning participants both in their communities and in the direction of their lives.

Inner Transition

As the Transition movement seeks a more community-based, ecologically connected, self-sufficient world, it recognizes that the crux of that change must happen within the heart and soul of the people. We need to discover our humanity—what it means to be human beyond our activities. We are not just workers or consumers, we are deeply caring individuals who love our children, our friends, our place in nature. But our culture celebrates the doing, the winning, the getting, and we forget the joy of being and sharing. That is why inner awareness or "inner transition" practices are integrated into every level of Transition Towns work, from self-reflection, to healthy group processes at a local level, to the integration of active listening in decision-making processes at the national and international levels.

Sophy Banks, founder of the first Inner Transition group in Totnes, proposes a task-relationships-process triangle for groups, where equal time is given to the project at hand (the task), how people are feeling about the way it is working (relationships), and what to do (the process) to improve the health of both the relationships and the task. Banks recommends that groups begin meetings with free sharing of each person's news of the day—happy or worrisome—and talk about ways to recognize and accept what is going on inside. That helps the individual to acknowledge stresses and anxieties, and it helps the listeners to be more receptive to that individual. Some disciplines suggest that as stress responses travel quickly from the brain through the nervous system to the body, if people can learn to recognize the feeling and cause of a physical response—the tight throat, clenched gut, speeding heart—they can head off the emotional reaction, and eventually learn to avoid the negative response entirely.

Groups may begin meetings with three keepers: a record keeper, a keeper of the allotted and used time, and a keeper of the emotional climate. If the climate keeper notes signs of dis-ease, she stops the meeting and the group discusses how and why it went off track and what can be done about it. Often an unhealthy atmosphere has to do with a member not understanding how to speak without sounding accusatory or how to listen without feeling defensive. Much effort is given learning how to listen and truly understand where another person is coming from.

In any healthy ecosystem, permaculture garden or transition meeting, diversity should be welcomed. Controversy is a good thing. It is essential to see all sides. A broad view makes a real solution more likely. Groups begin with ground rules as to how ideas might be presented or received, but the guiding precept is to welcome, listen, respect, and seek to understand. Carolyne Stayton, Transition US Executive Director, suggests addressing conflict by transforming it into a learning opportunity and pathway to personal growth as well as potential enrichment to the discussion at hand.

A similar approach is helpful for the individual—attempting to understand and listen without judgment to emotions that we might otherwise see as shameful or inadequate. People are also encouraged to be aware of extremes. Is the competent hardworking leader dominating, or does she empower others to use their talents? Sophy says a common necessity in functioning groups is constant decapitation of the ego. *It's not all about me.* Group members must focus on the goal.

Banks points out that the way we grow up helps form our worldview and our reactions, and it is those beliefs and narratives that create both the problems and the potentials of societies. As products of our childhood cultures, we likely hold many of their same old beliefs, even if those beliefs are counter to the world we want to see. She posits that one reason we have group relationship problems is that, while Transition seeks change toward an egalitarian society, most of us grew up under hierarchical models and tend to see ourselves either as leaders or followers rather than as equals. But agreeing together on group structure and process builds a functioning culture where people trust each other's motives and responses. Then each member should be able to listen to the "climate keeper's" report without blame or shame.

People problems can be internal as well as between members of the group. Banks says that our society shows the characteristics of bipolar disorder, cycling between manic activity and depression. The economic model demanding constant growth is fed by our compulsion for nonstop doing and going, speculating and buying. We are crazy-busy, activity squeezed in between activities, rushing wildly from place to place. Then we seek respite and distraction with television or smartphones or mood-altering activities to keep the feelings at bay that we've pushed away as we rush around.

Transition groups are not immune to the compulsion to overdo. The goals are important. There's work to be done. And in the case of climate change, there's no time to waste. But exhaustion and understandable despair do not help bring forth a new world. And although it takes a critical mass to effect change, recruits are not gained by frantic demands. The movement does grow, however, when it presents warmth and compassion.

To address this conundrum, Transition leaders urge groups not to exhaust themselves with the project at hand—the "doing," important as it may be to get done—but that they make time for "being," and for checking in with themselves and each other as to how they are feeling. Spending time feeling the breezes, listening to the birds, breathing in the smells of new leaf buds, moist soil, spring flowers, and taking time to just "be" feeds the spirit. Amidst world worry and social despair, it is also crucial to pause, and consciously venerate the things and people of the world held dear. Social time is important as well: having fun together without needing to be productive, and feeling part of a happy, empathic group.

The key to a healthy group is not in trying to make every moment smooth and upbeat, or to expect every individual to be perpetually calm and understanding. Rather, it is in finding ways to welcome divergent ideas and needs, and both to recognize un-health and remedy it. Marissa says, "Transition is all about relationships and mutual support: we believe connected communities are the foundation of the social change and ecological resilience we need in order to survive as humanity on this planet."

Mama Marissa

There's a picture of Marissa at the podium welcoming attendees to the big Transition US National Gathering, Summer 2017. In her arms is Gabriel, a chubby six-month-old gazing transfixed up into his mother's eyes.

I checked in with her in late spring, 2018. Gabriel is nearly a year and a half old now, and Marissa says things are going well in Paonia. Jeremiah, along with some other local growers, is starting an eighteen-acre organic hemp farm for producing CBD (cannabidiol), a compound with great promise as an anticonvulsant and anti-inflammatory product, as well as potential for combating anxiety, joint pain, insomnia, and more.

Marissa, with Gabriel's energetic help, is mobilizing families against fracking development in Colorado's Delta County. She calls this Op-

eration Mama Bear. She is also starting two "REconomy"-aligned local businesses: a sustainable event-planning business called "Mountain
Mamas Events"—partnering with another mama to host farm-to-table
weddings, retreats, and other gatherings; and "Paonia Home & Body,"
a worker-owned cooperative manufacturing artisan nontoxic/natural

Marissa, Jeremiah, and Gabriel at home in the North Fork Valley, Colorado.
Photo courtesy of Marissa Mommaerts.

cleaning products and toiletries (starting with laundry soap). These businesses, along with Jeremiah's farm, help localize the economy, serve people and the planet, and increase community resilience.

As if that doesn't keep Marissa busy enough, she's also still working with Transition US, building on the momentum from the 2017 national gathering to support more regional and national-level organizing, including the launch of several working groups focused on economic transformation, social justice, and inner resilience.

Regeneration

In their pre-parent days, Marissa and Jeremiah had planned to travel the country working on regenerative agriculture projects, and they spent their first Valentine's Day gardening. Working with the soil is basic to them as individuals and as a couple, but growing food has become an increasingly rare talent the last fifty years or so, as mega-corporations bought up land once owned by millions of US families. Nowadays, many folks are completely disconnected from the source of their food. Other living beings learn the business of finding nutriment before they leave the nest or their mother's care or their next molt. Humans *used* to know how to feed themselves until they left or were forced from their lands and gained conveniences that made them dependent.

I grew up in rural western Washington when kids (at least rural kids) were still taught how to grow and prepare food. During World War II, the federal government inspired a broad segment of the population to grow their own. "Victory Gardens" made food more accessible when so many agricultural workers were away at war and gas for transporting produce was rationed. Victory Gardens decreased the price of feeding the troops, and made average citizens feel as if they were contributing. By May 1943, eighteen million Victory Gardens grew in the United States, twelve million of them in cities. That was a garden or orchard each for about half the families then in the nation, together producing as much as the nation's total commercial output. The Executive Branch talked up the importance of home gardening, and the Department of Agriculture offered how-to information on growing and preserving. Though many people felt they were doing their bit for patriotism, more than half the people polled said they were growing their own for economic reasons. Now in the twenty-first century, people are beginning once again to appreciate a garden's potential. Marissa and Jeremiah's little Gabriel, all tan and gold from time outside, already knows how to hold the hose to help water the plants. He will know how to find or grow his own food long before time to leave the nest.

Marissa, along with many others, believes that regenerative agriculture may be the key to regain both the people's and the planet's health. Learning how to garden and to prepare and preserve food are clearly important steps toward food resilience and security. Gardening organically ensures that the eater will not be ingesting pesticides and other unwanted chemicals. But beyond those personal benefits, regenerative gardening—including any form of gardening (organic, permaculture, eco agriculture) that considers the soil community, biodiversity, and water conservation as well as productivity—provides for future generations as it rebuilds healthy soil ecosystems. As long as regenerative farming takes place widely enough that people have food access either in their backyards or from local farms, or have access to urban community gardens or subsidized markets and cafés, as in the Brazil example, the capacity exists to provide food for everyone.

Significantly, as people are being fed, so is the soil. Soil was once considered a renewable resource as it is indeed always being replenished. But it takes about a thousand years for natural processes to provide three centimeters of soil. Soil is currently being degraded or eroded away at a rate ten times (in the United States) to thirty or forty times (China) the rate of renewal. Each minute the world loses thirty soccer fields' worth of soil, totaling about twenty-two billion tons of fertile soil per year. If the world continues business as usual, no arable land—the soil necessary for growing food—will remain in sixty years, according to UN Food and Agriculture Organization Deputy Director General Maria Helena Semedo. Those business-as-usual soil-destructive practices include deforestation, over-grazing, and industrial agriculture mismanagement.

Hand in hand with loss of topsoil is the fact that agriculture uses 70 percent of the world's fresh water, and drought conditions are becoming increasingly common, causing hunger, emigration, and war. Degraded soil holds less than half the water of healthy organic soil. Swales built with the topographic contours allow rainwater to seep into the soil rather than run off. Organically rich regenerated soil builds the soil ecosystem. Soil microbes recycle organic matter, building soil structure to hold water like a good sponge, making that water available to plant roots.

Regenerative farming incorporates organic matter, uses in-farm fertility, rejects pesticides, minimizes tilling, and maximizes diversity, feeding the soil, bringing it back to life, helping it retain water and nutrients, and improving cleanliness of waterways. In a healthy functioning ecosystem, vegetation pulls carbon out of the air to feed the plants and stores that carbon in the ground. And therein holds another promise.

Because carbon dioxide remains in the atmosphere so long, the current level is sufficiently high that even if we changed to 100 percent renewable energy today, it could still trigger devastating warming. At this point, we must *remove* CO_2 from the atmosphere *as well as* rapidly cut emissions.

Earth-generated carbon dioxide is stored in three places: the atmosphere, the oceans, and the soil. The atmosphere is already supersaturated, warming the climate. Excessive carbon is acidifying the seas, dissolving the shells of sea creatures. But soils have capacity to store much more than now present. Estimates are that the soil has lost 50 percent of its stored carbon, due to harmful farming practices. Both soil and climate will improve by getting that carbon back in the ground. Regenerative agriculture has the potential to do exactly that.

Current studies indicate that if regenerative agriculture were widely adopted, from 40 to 100 percent of annual CO_2 emissions could be sequestered. Wide adoption might require something like the Victory Garden movement of the early 1940s war years. At the very least, government might shift subsidies from corporate agriculture to organic or other restorative forms of farming. It could provide education and "bully pulpit" inspiration. Government could impose a carbon cap, where a limit on emissions becomes stricter through time. The cap makes pollution expensive, which stimulates more energy-efficient technologies. Local gardening and farming then become more economical than buying store trucked-in produce. One way to help that government shift is to elect people to office who understand both the concept and the urgency.

A strong push toward regenerative agriculture could help bring the climate back from the brink of disaster, assuming renewable energy was concurrently replacing fossil fuels, and scaling up numbers of small local farms limits the controlling power of mega-corporations. But here again, the other necessary actions must be taken—all of the avenues need to work together. The populace must be willing to stop dumping excess greenhouse gases (or plastic or anything else) out into the air or water or soil. No more fossil fuel burning. No more exploitation (of resources or people). No more confined animal feeding operations. As living standards rise, as is happening in China, people want to eat more meat. Meat production uses much more land, calorie for calorie, than does vegetable production. And one third of agricultural land is used for animal feed rather than for human food. People can learn to eat less meat, plus choose animal sources that find their own food, as opposed to having acreage devoted to food grown for them. People might discover joys that exceed those of acquiring, therefore using fewer resources, and might opt for smaller families, using even fewer.

The promise of the Transition movement is its broad and deep perspective. It examines the many aspects of current western lifestyle harmful to society and to the environment, working with diverse voices to find effective, equitable local solutions. But to put it all together, the first and perhaps most important piece is inner transition. Changing course requires a change in attitude. People need to care more about each other and a viable future than they do about "winning," or about storing up maximum goodies in their short lives. And they need to care about listening to people of different backgrounds and cultures, because diverse insights help define problems. Real change will happen only when, as we saw in "From Conflict to Collaboration," the first story in this book, we find equitable common ground. Then together we can work toward a healthy biosphere and rewarding lives for the world's children's children and for all earthlings.

Based on permaculture principles as imagined by those students in County Cork, Ireland, and developed by Rob Hopkins and Naresh Giangrande in Totnes, UK, the Transition movement weaves together the diverse and essential parts of society, just as a healthy ecosystem is a product of all of its complex and diverse parts. Too often, good people feel the need to differentiate between these parts to choose what to them is the most urgent problem. To treat a symptom, so to speak, rather than the source of the disease. In fact, climate change exacerbates all of the other problems—food and water shortage, inequities, immigration issues, species loss, devastating weather events, war. And climate change, along with those other concerns, has at its base the unquenchable thirst for growth, for resource exploitation, for personal power.

Transition's essence is its inclusiveness and systems-thinking. Respecting the parts of self, of humanity, of earth's life. Recognizing people's health and happiness now and in the future as overarching considerations. Understanding the cause of problems and addressing them at their root. Learning to appreciate diverse points of view and working toward common goals.

Cowboy Gabriel with striped hen buddy. Jeremiah's hemp is visible in the background.
Photo courtesy of Marissa Mommaerts.

Together people can learn to farm sustainably. They can learn how to provide locally and equitably for their needs and for those of others. They can experience the joys of self-sufficiency, creativity, working in community, belonging. And then perhaps they can build the world they want to see. Together.

Conclusion

I SET OFF ON THIS EXPLORATION to find and tell stories of people working together for justice and the environment. I found so much more. I found depths of love for the land, for natural systems, for each other. I saw the boundless heart of humanity that I believe, together, is capable of redirecting our orientation toward one of a healthier and more just planet. People in these stories demonstrate choices beyond denial or distress. They have joined with others to help build the world they want to see and, concurrently, are themselves motivated and energized by their mission. Their work and spirit gave—and gives—me hope.

With squabbling ubiquitous in the news, whether between political parties or nations or beliefs, how refreshing it was to visit the Blue Mountains of eastern Oregon and watch conservation and logging interests learning to communicate, to talk and listen from the heart. How enlightening to see a mediator facilitate that communication. To see trust develop, to watch unfamiliar ideas understood and friendships bloom. Compassionate listening is, I believe, skill number one to working together for a better world, but it's a skill that is often misunderstood. It doesn't mean compromising your principles. It doesn't ask us to speak with one voice. It doesn't require finding a middle-of-the-road, namby-pamby, gutless non-solution that offends everyone equally. Rather, it honors diversity, seeks to understand, and focuses on how common goals can be sought, even though seen through different lenses. The

Blue Mountains Forest Partners' inspirational story is number one, "From Conflict to Collaboration."

What impressed me most in "Community," the story of Colorado's North Fork Valley (beyond the beautiful valley itself), was the love and dedication the citizens had for their home grounds. Many of them had *chosen* the valley—as opposed to just ending up there—as where they wanted to spend their lives. They would not easily let it be destroyed. That connection to the living world and sense of place is increasingly rare and precious. In the last several decades, our busyness and isolation, our cars and electronic gadgets, have pulled us away from wild nature. By 2014, 54 percent of the world's population lived in cities. In North America, urbanites were 82 percent of the population. As we move from rural life, it is easy to forget that we are part of the natural world. We are torn, or tear ourselves, from our roots—our moorings—our lifeblood. But still, most can find a place to claim—a piece of the sky, a window box, a city park or vacant lot, a street tree or crack in the sidewalk—a piece of the natural world. We can tune in to it like the people of the North Fork Valley tune in to their home. Pay attention. Connect. Hold it in our hands, our minds, our hearts. Cherish and protect: our environmental home needs our protection, and we need its gifts.

Water, air, food, and shelter received from the natural world are prerequisites for our survival. Less tangible, but clearly essential, are gifts such as reduced stress; improved cardiovascular health; enhanced concentration, memory, and energy; and the lifting of depression, according to studies cited by Qing Li, Associate Professor at Tokyo's Nippon Medical School and chairman of the Japanese Society for Forest Medicine (2018). Those measurable effects make me wonder how much urbanization and isolation from nature contributes to recent spikes in US addictions, depression, and suicides. E. O. Wilson (2009) says that because the human species has spent most of its history in the natural world, we are hard-wired to experience it. Just as our health improves from time spent outside, it suffers without it.

Sometimes we're so focused on our daily routines and anxieties that we forget to pay attention to the world beyond our immediate bubble. But becoming aware of our natural surroundings not only feeds our own well-being, it directs us to ways of helping each other: If pollution blocks the sky, paving covers the land, and the river is clogged with trash and toxins, that reveals a clear focus for people power. Qing Li lives in Tokyo, the world's most crowded city. He points out that the city's approximately 13.5 million people live in 2,191 square kilometers, which would average more than six thousand people packed into each square kilometer. But there are parks, urban forests, street trees. Storms and war devastated much of Tokyo's green spaces early in the twentieth century, but the city began a concerted effort to replant, and has recently been listed as having the largest tree cover per person of any megacity. Pressing our cities to provide pocket parks, street trees, community gardens, and green roofs could improve the health of both the populace and the climate. Spending time in those green spaces could save our sanity as well as our lives, E. O. Wilson and Qing Li's studies agree.

As I inwardly thank Drs. Wilson and Qing Li, my attention is snatched away by flashes of gold and black, of red and brown, in swooping, galloping, chittering flight, bursting by dozens from the feeder to the nearby blue blossom *Ceanothus*, back to the feeder to flutter, squabble, and line up expectantly along a branch, then explode once again to loop through the sky toward waiting oaks and Douglas firs. I think how passionately I want the finches to keep flying, the trees and shrubs to stay healthy, my grandchildren and great-grandchildren to delight in all this when they are my age. I want, as do the people of the North Fork Valley, to fight for the health of this planet every remaining day of my life.

For "Borrowed from Our Children," I was fortunate to discover the concept of the atmosphere as a trust right in my own hometown of Eugene, Oregon. Talking with environmental law professor Mary Wood and reading her enlightening book, *Nature's Trust* (2014), then listening to the arguments of attorney Julia Olson and others during court cases, brought home the pure logic of the Public Trust Doctrine. What could be more basic than a government's duty to protect the environment that we all require for life? And how very life-affirming is the doctrine of intergenerational responsibility—a long-held principle of Indigenous people, that today's decisions should respect consequences seven generations in the future. I was deeply moved also by learning of the dedication of countless attorneys across the nation and the seas, giving incalculable hours of time and energy with minimal compensation. I found their commitment at once stunning, humbling, and inspiring. May we all care more about life and the earth's future than about our own pocketbooks.

A lasting takeaway from listening to the "climate kids," from reading about the innumerable youth groups forming in our nation and around the world, and from talking with local high school students was a firm belief that if we accept our responsibility and rally enough people to get quickly on a planet-saving track, the kids will salvage more than we can imagine. Don't believe the things you hear about this generation being oblivious and buried in their electronics. Undoubtedly some are (as are some of their elders). But the young people I talked with, watched, and read about are savvy, passionate, informed, and active. The kids don't seem so weighed down with the baggage carried by most of their seniors. They don't have a portfolio they're afraid to lose, they're not hopelessly acclimated to the profit motive or to hierarchy, they're not inclined to stereotype or distrust those unlike themselves, and they're determined and focused. Beyond all the ever-growing numbers of climate groups, look at youth for sensible gun regulations. And Sweden's teen, Greta Thunberg, who, inspired by the strikes of the Parkland students, has in turn motivated students worldwide to demand that their elders wake up, take responsibility for bad decisions, and make an immediate U-turn.

I am reminded of a comment by plaintiff Hazel Van Ummersen, eleven years old in 2015, when she first became involved in the youth climate case, that she didn't see herself as a "climate kid." She relates more to the term "climate warrior." A growing number of this generation, the climate warriors, has had enough damage to their rights, their psyches, their futures. They're demonstrating. They're striking. And soon they'll be voting. I was incredibly moved by these young folks. The comments of two of the climate plaintiffs hang with me. Vic Barrett pointed out the importance of enlisting people from the margins of society to fight those powerful forces determined to maintain their pursuit of profit above all. People on the edges are particularly resilient, Barrett explained. They have had to learn to be tough "in a society not made for them."

Plaintiff Kiran Oommen knows the kids may not win, but the fight gives him a "more purposeful now" and the strength of community. The wisdom and perseverance of these young people is staggering. With such representatives of their generation in charge, perhaps humanity can learn that interconnection of all things includes us.

I traveled to the southern Oregon coast to learn more about the Jordan Cove liquefied natural gas export facility and pipeline ("Converging on the Cove") and was particularly impressed with the wide-ranging outreach and coordination of the groups opposing the fracked-gas energy project. What I've seen in the past is blocs working in separate silos, oblivious to, or even competitive with, other groups having similar goals. But the associations I found not only respect and seek diversity within their own groups, they welcome diverse outside groups to join in work for a united cause. Research and history show that a majority is not necessary to change a culture. Critical percentages for change range from at least 10 to 25 percent. But research is consistent in showing that the minority must be committed, aware of each other, and not squabbling within its ranks. These southern Oregon groups seem to be honoring all of those requirements: commitment, respecting diversity, and maintaining communication. Such networking and collaboration are keys to changing operating systems from profit and growth to honoring people and the planet. One of the great revelations of my research is that this collaboration appears to be exactly what is happening.

And I was constantly reminded of interrelationships. As I followed Marissa Mommaerts from anxious teen through international discoveries and committed work for resilient living in the fifth story, I was delighted to find her and her new family arriving in Colorado's Paonia, in the very North Fork Valley of story two. And then realizing the outsized role the Jordan Cove LNG export terminal and connector pipeline from story four would play in the North Fork Valley's fight against fracking underscored the importance in denying Pembina's permit. Interweaving becomes one of our most far-reaching and essential to remember concepts.

The Transition movement (story five, "Getting It Together, Together") looks at the numerous problems we face today and considers ways to bring forth healthier, more resilient, and happier lives. Its vision is that we can learn to take care of our needs and of each other, using permaculture principles as our guide. Derived from traditional societies that have lived long and harmoniously with their land, permaculture's guiding ethics of earth care, people care, and fair share would serve our human communities well. Without earth care, we have no home and no means of survival. Care of the soil helps sequester carbon as it makes possible the production of food for people. As the lungs and veins of the planet, forests and waterways must be kept healthy. The function of earth's systems and organisms requires respect. Earth care is ethic number one.

Part of earth care involves regenerating damaged systems. Improving the soil brings the subsurface ecosystems to life and pulls excess carbon out of the atmosphere. Closer to home, growing and preserving one's own food is a ticket to freedom, thus also qualifying as people care. Civil rights organizer Fannie Lou Hamer famously said, "If you have four hundred quarts of greens and gumbo soup canned for the winter, no one can push you around or tell you what to do."

Transition considers diversity and opportunity, food and health as part of people care, but to work for those things requires inner transition. As Mahatma Gandhi said in 1913, "All the tendencies present in the outer world are to be found in the world of our body. If we could change ourselves, the tendencies in the world would also change."

Transition work integrates "inner transition" into every level, explains Carolyne Stayton, Transition US Executive Director: from self-reflection practices at a personal level, to healthy group processes at a local level, to the integration of active listening in decision-making processes at the national and international levels.

Sharing with each other how each person *really* feels (beyond "Fine, thank you") bridges differences and facilitates trust. I well remember learning that lesson myself. My family was loving, but reserved. We were slow to complain and certainly didn't "hang our dirty wash in public" as my mother called airing troubles. Thus, though I had many friends, the friendships weren't as deep as they might have been: I tended not to share my inner feelings. But I discovered in my forties, as I was training for marathons with friends, that running loosens protective armor as well as the joints. As a friend speaks from her heart, the listener feels trusted and in turn dares trust. It is a wonderful, liberating, loving feeling. Stayton says she gets to "see the precious core" of her colleagues and in turn be "deeply seen. There is nothing quite so fulfilling."

Sophy Banks, Transition Network Inner Transition Coordinator, speaks of a conscious, compassionate, and spiritual change. The first step is to imagine the kind of future we'd like to have. In so doing, we must acknowledge that our present western society is based on individual separation, division, and competition. We must understand how that manifests personally, and how we would like it to change both in ourselves and in the world. Once we stop trying to preserve the image we or our family or the TV tells us is presentable to the public and accept ourselves in all our flawed glory, layers of stress peel away. Like a veil is removed, we're able then also to see the humanity in others.

Recently, I was browsing a writing magazine (*The Writer's Chronicle*, 2019) and read an article about expanding character in fiction through exploring the vulnerability exposed by the character's outrage. It occurred to me that we could gain the same insight in life by considering the basis for the passions of people with whom we disagree. From that could come understanding, the first step in finding mutuality and perhaps the "precious core" of another person. Together we could then create a new story, one based on us—all of us, not just me. We could ask, what do we all need, as individuals and as communities? How can we work together toward mutual goals?

That shared humanity can grow through ways of being, rather than just doing, a spiritual state Sophy says we've largely lost. We sometimes experience that state through music or dance or in communion with the natural world. Sometimes in deep closeness with other people. I feel grateful to have tasted such perspectives with friends, through singing heartily with other voices, and in my own green world. One time I was sitting in our woods, admiring the way the low sun sparkled the dew drops on long loops of lichen, and I suddenly found myself in tears. Inexplicably, I was envisioning starving children in an unidentifiable impoverished country. There is nothing rational about that, but it was powerful and real and unforgettable. Another "irrational," and I would say spiritual, experience was when a friend gently pressed my husband's temples with *ki* and (as far as I'm concerned) saved his life in the midst of a brain bleed, an otherworldly experience and source of lasting gratitude.

Once open to other ways of knowing, it is far easier to listen deeply to others, perhaps because ideas are no longer so regimented, so black and white. And when people are able to find a receptive and empathic outlook, they are more able to experience compassion not only for the human species, but for all earthlings and the systems they require.

Basic to Transition is that compassion for all parts of the biosphere, an attitude that appears essential to continuing maintenance of living communities. I recently read some speculative history of Indigenous Pueblo tribes in New Mexico's Tewa Basin (Langlois, 10/2/2017), demonstrating efficacy of cooperative over competitive community. It appears that many, if not all, of the people who established the resilient community on the banks of the Rio Grande in the late thirteenth century migrated from the ancient Mesa Verde cliff dwellings in southwest Colorado. Between 900 and 1200 CE, Mesa Verde's population tripled. Each family had its own garden and other belongings in this agrarian society, and the population explosion bred haves and have-nots, bringing predictable acrimony. Evidence shows a sudden exodus from Mesa Verde in the late 1200s, coinciding with signs of violence. Though there currently is no proof of the journey taken, seeking the life-giving water of a river would have been paramount. A large community by the Rio Grande near present-day Santa Fe, the Tewa Pueblos bloomed shortly after the exodus from Mesa Verde, and its handed-down stories reflect a comparable long walk. The Tewa Pueblos have maintained a resilient, functioning, communal society for the past nearly eight hundred years.

The river, of course, has made a difference in the health and resilience of the community, but the existence of springs near Mesa Verde show that lack of water was an unlikely cause of the demise of that population. The biggest distinction between the longevity of the two sites appears to be the cooperative culture in the Tewa Basin versus the individualistic culture in Mesa Verde. The Tewa Basin Pueblos have cooperative gardens and gardening, cooperative harvest and sharing of harvests, and large plazas to gather, discuss, plan, and celebrate together. Two divergent states of mind, or perhaps states of spirit—the scarcity, get-it-quick-before-it's-gone spirit as opposed to the grateful spirit of abundance: together we honor the good earth and share its products.

Transition groups are finding ways to live well, respecting and working cooperatively with others and honoring and helping to renew natural systems depleted by extractive practices. Caretaking rather than consuming requires a sharp turn from the acquisitive, competitive individualism that western society tends to encourage.

Underlying the challenges in *Shoulder to Shoulder*'s stories is the age-old dichotomy between democracy and plutocracy—the needs of the people versus the entitlement and power of extreme wealth. In the United States, that struggle is as old as the earliest colonies. Wealthy colonists, primarily British, came to America to escape the monarchy, not to establish justice and equality for all. Voting was originally only by (white male) property owners, about 10 percent of the population.

Today we have come far with voting rights, but as Supreme Court Justice William J. Brennan once said, "liberty is a fragile thing." We need always to be on guard. And the creeping influence of moneyed interests goes deeper than many people are aware. Currently, there are vested interests in every realm of public policy: international trade, the environment, defense, agriculture, health care, you name it. With money's fingers in the pies, the pies are made to money's specifications.

We have a consumer culture that expects businesses to profit and the economy to expand at whatever cost. The system assumes winners and losers. Collateral damage is expected and accepted. Growth is required; resources are the means. But resources expand slowly if at all. The essence of water, air, forests, soil, minerals, fossil fuels, and people diminishes as they are polluted, abused, or disrespected. Each has its own intrinsic value, and to exploit it is to exploit the function of the systems that all depend on. #Metoo didn't go far enough. Hooray for the cultural wake-up call that women are not merely resources. But neither are the struggling workers, female or male, mere resources. Neither are those replaced by machines. Neither are the ecosystems or their integral parts. Such exploitation is the basis for economic inequality. It is the basis for climate change. It is the basis for species extinction. The idea that any individual could be considered superior to others, and therefore deserving of usurping, colonizing, exploiting other beings, or resources—portions of the commons—is the crime, and is fundamental to both societal and environmental problems.

Thus many legal changes are required, historically accomplished only with overwhelming public pressure. In many ways, this period of history, the end of the second decade of the twenty-first century, seems a very dangerous time. Inequality is at its highest in the last fifty years, freedom of the press and assembly are threatened, corporate rights are favored over those of the environmental commons or of common humans. But in other ways, now is the most hopeful and exciting time in that half century as increasing numbers of people awaken to those dangers and focus more clearly on vital values. People were cautiously hopeful after the Paris Accord. Though many knew it was inadequate, most were thrilled at a step in the right direction. But as the new US administration reverted to a regressive, anti-Paris agreement, more extreme profiteering model, throngs cycled through shock and despair, then awoke one day to resolve: *This will not abide!* Masses of people want change and are pushing for it to happen. Witness the women's march, the climate march, the march for science, the amazing Greta Thunberg and the kids' school strikes, the Sunrise Movement, the United Kingdom's Extinction Rebellion. Midterm elections in 2018 featured uncommon organization for progressives, increasing diversity and democracy. The old disorder is challenged when people come together.

Changing a system and its accepted goals and mores seems daunting. Yet in my lifetime I have seen more than one such broad social change. My young childhood saw mutual trust as jobless strangers split piles of wood in exchange for hot meals, and people routinely gave or hitched rides with people they didn't know. That kind of trust is a rare commodity now. I believe the mistrust is largely manufactured, stemming primarily from interests profiting from a divided and fearful populace, but it can probably be attributed also to population increases and urbanization. Whatever the cause, divisiveness has been a striking cultural change.

Another enormous change was the advent of the chemical age, a change with strikingly mixed blessings. During and before my young childhood, food was grown organically: farmers fertilized with compost and animal manure and fought garden pests like my mother did, saying that a gardener's green thumb was earned by squishing aphids, or in more problematic ways, like spraying with lead arsenate. Insect pests troubled not only farm products; they also were powerful vectors of disease. Throughout all previous major wars, more people died of diseases, like insect-carried typhus and malaria, than from bullets. But during the prelude to the Second World War, a Swiss chemist discovered that DDT would kill those insects. Because of this miracle chemical, military deaths due to disease were dramatically decreased. And DDT ushered in a new wave of chemical discoveries and magic bullets in agriculture against both insects and weeds, the cattle getting fatter and crop volumes higher. Eventually, scientists discovered that DDT accumulated in predatory birds, various marine animals, and humans, and by 1972 was banned. But today pesticides pollute waterways, kill pollinator insects and the soil microbiota.

Though the simpler living and trust of the mid-1930s to '50s have largely disappeared, positive social changes have happened as well. I didn't see racial segregation in the small lily-white Pacific Northwestern town where I grew up, but in the south Blacks and Whites were relegated to separate swimming pools and drinking fountains. And when I came to college at the University of Oregon in the mid-1950s, a cross was burned on a sorority lawn because a sister, later a friend of mine, was dating a Black student. There was also at least one restaurant in town where that couple was not welcome. The Civil Rights Act of 1964 was won during my young motherhood, outlawing such egregious practices. Women have made strides as well. Still socially and often legally subservient to their husbands in my growing-up years, in 2019 at least five smart, independent women are running for president.

Attitudes can change and so can laws. Now though, in order to make the changes necessary for the health of the biosphere, we must make major upgrades to our basic operating systems and make them quickly. Some contend that's impossible. The suggested changes are too great, they say. Our current way is too ingrained. Change would destroy our economy. We the people wouldn't be willing to make the necessary sacrifices.

I don't believe it. I've not only seen major changes in social attitudes in just a few decades, I see countless groups and individuals eager to do what is necessary to further fairness, respect, and health of people left behind, of other species, of ecosystems. The Transition movement, 350, and the Movement Generation are just three such groups. The Movement Generation Justice and Ecology Project, centered in Oakland, California, is dedicated to the "liberation and restoration of land, labor, and culture." Focused particularly on low-income communities and communities of color, it is committed to a transition "away from profit and pollution and toward healthy, resilient, and life-affirming local economies." Movement Generation, as well as numerous other groups, is rediscovering organic agriculture, natural health methods, and ways to regenerate soil systems.

Joanna Macy's "Work That Reconnects" helps people find connections with each other and with earth's life systems so they'll be motivated to do their bit toward sustainable civilization. And numerous other groups profiled on these pages and beyond see the need to dispense with business as usual and move toward a society whose bottom line is health of the living world in all of its forms and functions.

Naomi Klein (2014, pp. 450–64) reminds us that now is a challenging time to make necessary changes to our economic system. The ultra-wealthy enjoy fewer restraints, greater political and cultural power, and a lower tax base than anytime since the 1920s. They will not easily give up the position. But though challenging, it's not the first time our nation has had to face down powerful interests. Notably, there is precedent in the abolition movement for such extreme economic change. In the nineteenth century, US and South American commerce and wealth were largely dependent on slave labor, as pointed out by author Adam Hochschild in his *Bury the Chains* (2005). The global economy's dependence on fossil fuels is comparable. Changing to renewable energy, as addressing the climate crisis requires, will benefit the great majority of the populace as well as the climate. But it asks certain powerful political and economic interests to give up enormous wealth. Fossil fuel folks won't go quietly, but neither did slave owners or traders. It was the strength of abolitionists and their mass movement that turned the tide. A similarly strong movement is required now. And I believe it's happening.

Still, I worry that we will put too much faith in quick fixes: Just get the right people in office or the wrong people out; get the right party or system in play and all will be well. We can go back to our lives. Back to business as usual. But current situations don't allow us to tune out, and we must say goodbye to business as usual. Goodbye to acquisition. Goodbye to exceptionalism. Goodbye to growth unless it's the growth of heart and soul. Goodbye to magic bullets.

What we need more than a change of personnel, party, or system is change within ourselves. I must erase from my personal scoreboard the comparisons of wins or losses from what I perceive as "my side" and focus instead on the wins for life—for *our* side: the wins for the biosphere and for the human populations, especially those currently at risk. Widespread change in objectives and measures of success suggested by Sophy Banks are necessary to grow the movement and to keep it strong through the predictable blowback from interests of the super-wealthy. Let the goal be for food, health, and opportunity for all. For the people to reconnect with each other and their earth home. For competition to be with ourselves, to be the best we can be while helping others and earth's ecosystems to be their best as well.

If I were writing a prescription for joining the climate and justice movement, it would begin with a call to go outside. Daily, or as often as possible, with no agenda and no phone, sit and be. Attend. Breathe in the fragrances. Feel the breeze on your skin. Listen to the birds, the wind sowing in the trees, unseen animals moving through the forest duff. See, and swell with burgeoning awe and gratitude at being part of all this.

It's the natural world that keeps me steady. The biosphere I require for survival, the biosphere I am a part of, my home and my joy. But the natural world is in trouble. I am as compelled to do everything possible to save it as I would be to save a house full of children or animals in the path of a fire. I am sorely inadequate working alone, but soloing isn't necessary. Welcoming groups are growing exponentially, groups dedicated as we are to saving our home and each other. Groups where we can give back and look forward, shoulder to shoulder.

Epilogue

Checking in with Blue Mountains Forest Partners

A WARMING CLIMATE AND DECADES OF mismanagement turned western forests into tinderboxes. At the same time, rural communities dependent on the forests for jobs were seriously stressed. But in the eastern Oregon and Washington Blue Mountains, a collaboration of timber interests, the conservation community, local residents, the Forest Service, and elected officials seems to have woven an effective strategy. Since the beginning of the BMFP, unemployment in Grant and Harney Counties has dropped by half. John Shelk's Ochoco Lumber has added a refurbished small-diameter mill into the ground floor of its Malheur Lumber Company mill in John Day. It will allow processing of the smaller and more fire-prone logs that must be removed to improve a forest's resilience. The new mill will give a welcome destination for juniper, as reduction of this thirsty and flammable species has become of increasing importance. Heavy grazing and fire suppression allowed juniper forests to expand until they outcompete grasses and shrubs that provide habitat for wildlife such as sage grouse. They deplete groundwater and understory with concurrent erosion and they decrease stream flows and fish habitat. But with little market for juniper, removing it has been an expensive proposition. The new mill speaks to that.

Ochoco and Malheur Lumber Company president Bruce Daucsavage says that with the cooperation of the collaborative plus the mill improvements, the company has been able to positively affect 86 percent of private employment within Grant and Harney Counties. The collaborative model has now spread to twenty-seven Blue Mountain communities currently working to restore public lands and private employment. Dave Hannibal, base camp manager for Grayback Forestry, working with prescribed fire, small-scale logging, and reforestation on these projects, says that he has seen a meaningful change in the forest. Because of Grayback's work in the first project of the ten-year stewardship, "A fire could burn in there and it wouldn't be a very big deal." A gratifying observation for all of the people and wildlife counting on the forests for their futures.

Collaboration isn't magic, of course. Some people still feel left out; some still prefer defending their position to hearing others. Often financing is irregular and uncertain, and frequently the commercial part of the forest work takes precedence over the essential, but noncommercial, restorative work. Changing personnel and priorities of the Forest Service can throw a monkey wrench into the works. But many people on the ground are trying to do the right thing for both the forest and the human community, and positive results help the collaborative model to spread. It is a model that could spread far beyond forest management. Whatever their income or politics, everyone needs clean air, potable water, and a livable climate, just as we all need food and shelter. Following the BMFP model could help people work together toward common goals rather than focus on their differences.

The North Fork Valley

When the Bureau of Land Management's Uncompahgre Field Office released its long-awaited Resource Management Plan, the valley populace was dismayed to see how completely it reflected the new federal administration's focus on boosting domestic energy production. Rather than setting aside 177,700 acres for protection, as had been included in the BLM's draft plan, the agency's current management plan protects none. Further, it doesn't consider wildlife habitat or migration routes, and it allows an amazing 27 percent increase in greenhouse gas emissions.

Colorado's governor Jared Polis, in a "consistency review," listed those and other areas where the new RMP is inconsistent with state laws, plans, and policies that protect wildlife and air quality. In late February of 2020, PEER (Public Employees for Environmental Responsibility) found, by way of the Freedom of Information Act, documents showing that the abrupt turnaround in the Resource Management Plan was dictated by the federal BLM. The feds told the locals that their plan was not in line with the administration's directive to increase access to oil and gas. This from a federal agency professing to embrace local decision-making.

While Citizens for a Healthy Community and the rest of the valley are buoyed by the support of the governor and by new state legislation to protect the environment, they remain vigilant against the conflicting visions of an environmentally healthy valley or having their home industrialized with expanded fracking.

August 7, 2020, *The Washington Post* (Eilperin, 8/7/2020) reports that a 30,000-square-mile area incorporating portions of Colorado's Western Slope that includes the North Fork Valley, along with three nearby counties in eastern Utah, has warmed in excess of 2 degrees Celsius since the beginning of the industrial revolution. More than double the global average increase, this makes that Western Slope area the largest hot spot in the Lower 48, according to *The Washington Post* analysis. Such heat, along with a twenty-year drought, bodes poorly for an agricultural community, and makes even more inconceivable anyone's support for a water-guzzling industry whose very existence, as well as end product, emits more greenhouse gases to the atmosphere, furthering and hastening still more warming.

August 19, 2020, climate groups including CHC, Colorado Sierra Club, WildEarth Guardians, and others sue the Trump administration, saying that the Bureau of Land Management violated federal law by not considering in its 20-year management plan how more fossil fuel development could harm organic agriculture, the climate, and endangered species. Attorneys from Western Environmental Law Center and the Center for Biological Diversity cite the *Post* study that pointed to the drying of the Colorado River, which supports endangered fish, agriculture, and 40 million downstream water users.

Fossil fuel production on public lands causes about a quarter of the nation's greenhouse gas emissions. If developed, pollution from already leased claims would exhaust the US budget that limits warming to 1.5 degrees C. Existing laws give Congress and presidents the right to end fossil fuel leasing on federal land. We the people, together, can demand it.

#Youthvgov

January 17, 2020, a divided panel of the Ninth Circuit Court of Appeals recognizes the importance of the young people's injuries from a heating climate and that the government's promotion of fossil fuel use is pushing the nation toward collapse. But two of the three judges find against allowing the case to go forward, saying that the executive or legislative branches are where necessary remedies should be addressed. In a passionate dissent, Judge Josephine Staton writes, "Seeking to quash this suit, the government bluntly insists that it has the absolute and un-reviewable right to destroy the Nation . . . [and] My colleagues throw up their hands."

Seeing that in most aspects the judges had agreed with the plaintiffs, in spite of two of the three finding against their right to be heard, at-torneys for the plaintiffs will now file an *en banc* petition, requesting the full bench, or eleven of the twenty-nine active Ninth Circuit judges, to review the case.

Clearly, each setback has to take an emotional toll, but the plaintiffs and their attorneys are undeterred and work toward action around the nation and the globe. September 2, 2020, fifteen Mexican youth from the State of Baja, California, file against their federal government as cli-mate change threatens access to food, health, and water for millions of citizens. September 17, 2020, in the midst of wildfires that have burned more than half a million acres in Washington State, a masked Andrea Rodgers argues to three masked judges in a courtroom empty of ob-servers because of COVID-19, that climate change is not outside the scope of judicial review. As he listens to arguments, an empathetic Judge Mann, turning to the defendants, says, "For the last seven days I can't go outside. If I go outside, I'm threatening my life. I have asthma, so I have to stay inside with the windows shut. I don't have an air conditioner. Why isn't that affecting my life and my liberty?"

September 24, 2020, the David Suzuki Foundation and the Pacific Center for Law and Litigation host an online briefing to prepare the public for the next week's virtual hearing of fifteen youth suing Canada to protect their rights to a livable future.

Our Children's Trust daily gathers more supporters. When the non-profit organization began, the climate was little on the mind of most of the public. Now the movement is strong and growing among youth worldwide; more adults are tuned in to the emergency; law students are beginning to protest against firms representing fossil fuel companies and to urge fellow students not to work for those firms; and the climate has taken center stage amongst 2020 presidential candidates. The tide is turning.

Jordan Cove

In May of 2019, the Oregon Department of Environmental Quality finds that the proposed liquid natural gas export terminal and pipeline had not been able to meet Oregon's clean water standards, and there-fore it denies Pembina's clean water permit. January 21, 2020, the

Department of State Lands denies the project's request for yet another extension, saying that the company had still not provided requested information for several previous extensions. The DSL was to announce a final decision on the Canadian company's removal/fill permit by the end of the month. On January 23, Pembina withdraws its application for that permit, though the removal/fill permit is required for the company's planned dredging.

Landowners, tribal members, government agencies, recreationists, youth, and many more had submitted nearly 50,000 comments against the Jordon Cove project. Thousands attended meetings and lobbied in Salem. State agencies listened and did their own research. The project was simply not able to show its ability to perform to standards, and a large contingent of the public made sure that the government agencies were watching. The project's critics are cautiously hopeful that the threat has ended. Still, they remain aware that this zombie project tends not to stay dead. The Federal Energy Regulatory Commission was scheduled to decide February 13 whether the Jordan Cove Energy Project should proceed. As the clean-air and removal-fill permits are required by Oregon law, should FERC approve the application and Pembina argue that a federal okay preempts state law, the case would almost certainly end up in the courts. February 13 comes and goes with no FERC decision. February 19, Oregon's Department of Land Conservation and Development denies the project's permit, detailing its projected negative impacts. February 20, FERC commissioners vote 2-1 not to move the project forward, but leave it open to further analysis. Pembina has thirty days to appeal.

As the state, along with the rest of the nation and world, retreats to their homes in the midst of a global pandemic, federal agencies proceed full steam ahead in approving massive fossil fuel export projects like Jordan Cove. In mid March FERC gives Pembina approval to go ahead, conditioned on its getting the missing permits from Oregon. On the same day, Pembina files a request to the US Secretary of Commerce to override Oregon agencies' permit denials. With FERC's conditional approval, Pembina can begin using eminent domain to force landowners from their properties. Richard Glick, the dissenting vote on the 2-1 decision, tweets "FERC's shoot 1st & ask questions later attitude is problematic," referring to the potential of someone being evicted from their home for a project that may never happen.

Governor Brown, who had received criticism for maintaining a low profile on the project, says, "I will not stand for any attempt to ignore

Oregon's authority to protect public safety, health and the environment." She also refuses to allow any condemning or clearing of property before all permits are received, and makes clear that she is ready to make a legal challenge if necessary.

Meanwhile, Pembina has not backed off, but with the current extreme downturn in oil and gas markets and Asian prices too low for profitability, the passions to proceed have probably cooled. Or at least been pushed to a back burner.

Transition

Port Townsend's Local 20/20 thrives and expands. Its wide-spreading chaordic structure encourages many people to spearhead or join actions, and provides sufficient diversity and overlap to absorb natural rise and fall of the people's energies and interests. The city's stunning, somewhat secluded, coastal site and welcoming town with its vestiges of Victorian history make it a popular retirement destination, contributing to Jefferson County's being one of the "most aging" populations of any county in Washington State.

To remain locally resilient, 20/20 continually assesses the spectrum of needs for a community to live well now and into the future. Beyond providing food, water, energy, and jobs for current needs, resilience clearly requires engaging and retaining local youth. Local 20/20 responds with a growing number of intergenerational programs. It works with partners from the high school's Students for Sustainability in events such as the Repair Café and a Jobs/Career Fair, staged at the high school. Local businesses set up tables at the fair where they could describe their work, their wage scales, advancement possibilities, and opportunities for necessary training. The youth could see themselves as valuable to the community fabric and see possibilities for their future that didn't involve moving away. L20/20 also partners with Skillmation, a local web-based program that aims to "connect people seeking mentorship with those interested in providing it." Skillmation has recently worked with high schools to provide each ninth-grader a mentor, a great entry onto a path toward success. Local 20/20's interweaving, rather than top-down, structure and periodic reassessment of what is essential for a vibrant and lasting community undoubtedly contribute to the group's continuing health.

Some programs do discontinue. Bay Bucks founders went on to other projects, and the group disbanded. Marissa Mommaerts says that she has learned to be comfortable with such dissolutions, considering them experiments to be learned from that also make ripples in other places and other lives.

Transition US reports nearly two hundred Transition Initiatives throughout the nation, with dozens more looking into the idea every day. Its ultimate vision is for every community in the United States to unleash its "collective creativity" for greater resilience and justice. The TUS office guides individuals and groups throughout the nation in localized resilient living, organizes workshops, and provides handouts on everything from emergency preparedness to inner transition. But the group knows it can't cover all the bases alone, so it welcomes partners.

Recently, Transition US has joined a coalition of climate, youth, and Indigenous groups, including 350.org, Rainforest Action Network, Sunrise Project, Sierra Club, Greenpeace, Oil Change International, Center for International Environmental Law, Union of Concerned Scientists, and others to "Stop the Money Pipeline."

Fossil fuel projects could not happen without the help of financial institutions. In a study of thirty-three banks, Rainforest Action Network found that financiers provided $1.9 trillion to the industry in the two years between 2016 and 2018. Through a vigorous information and activist campaign from the Stop the Money Pipeline consortium as well as from other people-powered groups, financial companies are becoming increasingly aware of their roles in energy transition. Several global banks, including Standard Chartered, BNP Paribas, Royal Bank of Scotland, and JP Morgan, have committed to stop providing funds for new coal mines or coal-fired power stations.

Most European, Australian, and Canadian-headquartered banks have at least signed on to provide climate-related financial disclosures, but less than half of US institutions have done so, according to Lauren Compere, Director of Shareholder Engagement at Boston Common Asset Management, an employee-owned firm specializing in responsible investing. Activists in Stop the Money Pipeline are targeting the world's largest private funder of fossil fuels, JP Morgan Chase, the top fossil fuel insurer, Liberty Mutual, and BlackRock, the world's largest investment fund, with over $7 trillion under management in early 2020. In response, Liberty Mutual adopted a coal policy in December 2019, the eighteenth global insurer to restrict coal insurance. It's an insufficient step, certainly, but it's a step.

A larger step comes mid-January 2020 from Larry Fink, CEO of BlackRock. In a letter to investors citing BlackRock's own research as well as that of international scientific consortia, Fink emphasizes the financial and socioeconomic risks presented by climate change, "deepening our understanding of how climate risk will impact both our physical world and the global system that finances economic growth." He goes on to say that investors are recognizing that "climate risk is investment risk." This is a game-changer. Without investors, banks cannot lend money. Without loans, fossil fuel projects cannot proceed.

"Feels like there really is a new round of strength forming in these burgeoning collectives and coalitions," Carolyne Stayton, co-chair of Transition US, says. "Hallelujah!"

Once the financial industry realizes that fossil fuels are no longer a good investment, a dramatic turnaround will follow. Our power structures have gone seriously astray, threatening our fellow humans, other biotic communities, and the very systems giving us life. But together, we are making a difference and can absolutely make far more.

Transition US is in the midst of a year-long national campaign exploring most effective ways to come together to cultivate a just, sustainable, and regenerative future. The group calls the campaign "From What Is to What If," considering the heart's visioning, the head's strategizing, and the hands' action. At the same time, TUS is exploring new leadership possibilities. After eleven years of dedicated work, Carolyne Stayton is retiring and returning to Canada and family. Assuming the role of Interim Executive Director is Don Hall, who we met with his ingenious quadruple-win Suncoast Gleaning Project in Florida. Interim Assistant Executive Director is none other than Marissa Mommaerts, who we followed from Wisconsin to Washington, DC, then to Sebastopol, California, back to Wisconsin, to land in Paonia, Colorado, in the North Fork Valley of the Gunnison River. The future of Transition US looks bright.

The year 2020 has been a challenging one in an increasingly uneasy world, but that world is in our hands. Important movements, big changes, have always been the work of the people. In the words of the great lady, late Supreme Court Justice Ruth Bader Ginsburg, "Fight for the things you care about, but do it in a way that will lead others to join you."

Increasing numbers of people are finding active passionate groups committed to the biosphere and to each other. And increasing numbers of groups are coalescing for greater strength and understanding. It is time and past time that we seek out these groups and join shoulder to shoulder to reroute our cumbersome ship of state and set it on the path to a viable future.

Appendix

THE FOLLOWING ORGANIZATIONS appeared on these pages or were connected with those that did. Many other fine groups are not listed. For information, to join, or to support, see these or other websites.

350.org is the international group. Also see your local 350 chapter
Appalachian Voices—appvoices.org
Blue Mountains Biodiversity Project—https://www
    .bluemountainsbiodiversityproject.org
Blue Mountains Forest Partners—https://www
    .bluemountainsforestpartners.org
CALNG—citizensagainstlng.com, now a project of Citizens For
    Renewables
Cascadia Wildlands—cascwild.org
Center for Biological Diversity—https://biologicaldiversity.org
Citizens for a Healthy Community (CHC)—www.chc4you.org
City Repair—cityrepair.org
Climate Action Network International (CAN)—climatenetwork
    .org (a worldwide network of over 1,300 groups in 120
    countries) CAN also has regional network hubs
Climate Colorado——climatecolorado.org
Colorado Farm and Food Alliance—www.coloradofarmfood.org
    (many local groups)
Columbia Gorge Climate Action Network—CGCAN.org
Douglas County Global Warming Coalition—www.douglas
    countyglobalwarmingcoalition.com

Earth Guardians—earthguardians.org
Earth Justice—https://www.earthjustice.org
Edesia Community Commercial Kitchen—http://www
.edesiacommunitykitchen.com/
Elsewhere Studio—elsewherestudios.org
Hair on Fire Oregon—haironfireoregon.com
High Country Conservation Advocates—https://www.hccacb.org
Huerto de la Familia—huertodelafamilia.org
iMatter—imatteryouth.org
KS Wild—kswild.org
League of Women Voters—lwv.org
Local 20/20—https://l2020.org
Lopez Community Land Trust—https://www.lopezclt.org
Mountain West Strategies—mountainweststrategies.com
Movement Generation Justice and Ecology Project—
movementgeneration.org
National Family Farm Coalition—nffc.net
New Economy Coalition—neweconomy.net
Northwest Earth Institute, now EcoChallenge—ecochallenge.org
Northwest Environmental Defense Center (NEDC)—law.lclark
.edu
Oil Change International—priceofoil.org
Oregon Shores Conservation Coalition—oregonshores.org
Oregon Wild—oregonwild.org
Our Children's Trust—ourchildrenstrust.org
Post Carbon Institute—postcarbon.org
Public Lab—publiclab.org
Red Earth Descendants—www.redearthdescendants.org
Renew Oregon—reneworegon.org
Rogue Climate—rogueclimate.org
Rogue Riverkeeper—rogueriverkeeper.org
Sierra Club—https://www.sierraclub.org
Solar Energy International—https://www.solarenergy.org
Southern Oregon Climate Action Now—socan.eco
Sunrise Movement—www.sunrisemovement.org
Sustainable Northwest—sustainablenorthwest.org
The Endocrine Disruption Exchange (TEDX)—https://www
.endocrinedisruption.org
The Wilderness Society—wilderness.org
Transition US—transitionus.org

Waterkeeper Alliance—waterkeeper.org

Western Environmental Law Center (WELC)—https://www
.westernlaw.org

Western Slope Conservation Center, aka The Conservation Center
(TCC)—westernslopeconservation.org

WildEarth Guardians—https://www.wildearthguardians.org

Wilderness Workshop—wildernessworkshop.org

In a separate category, as we have been dealing with local organizations, but listed because of its essential and transformative work and because its home office is in my hometown:

Environmental Law Alliance Worldwide (ELAW) is a global alliance of attorneys, scientists, and other advocates that assists local efforts to build just and sustainable futures, founded by public interest lawyers from ten countries in 1989—elaw.org

Bibliography

Bishop, Ellen Morris, *In Search of Ancient Oregon: A Geological and Natural History* (Workman, 2006)

Colborn, Theo, Dianne Dumanoski, and John Peterson Myers, *Our Stolen Future: Are We Threatening Our Fertility, Intelligence, and Survival? A Scientific Detective Story* (Dutton, 1996)

De Vriend, Wim, *The Job Messiahs: How Government Destroys Our Prosperity and Our Freedoms to "Create Jobs"* (Golden Falls Publishing, 2011)

Eilperin, Juliet, "This giant hot spot is robbing the West of its water," *The Washington Post* (8/7/2020)

Ferguson, Cody, *This Is Our Land: Grassroots Environmentalism in the Late-Twentieth Century* (Rutgers University Press, 2015)

Gore, Al, screenplay, *An Inconvenient Truth* [documentary]. Davis Guggenheim, Director, Laurie David, Lawrence Bender, Scott Z. Burns, Producers (Laurence Bender Productions, 2006)

Hacker, Jacob S., and Paul Pierson, *Winner-Take-All Politics: How Washington Made the Rich Richer—and Turned Its Back on the Middle Class* (Simon & Schuster, 2010)

Hawken, Paul, *Blessed Unrest: How the Largest Movement in the World Came into Being and Why No One Saw It Coming* (Viking, 2007)

Hayes, Christopher, "The New Abolitionism," *The Nation* (4/22/2014)

Hochschild, Adam, *Bury the Chains: Prophets and Rebels in the Fight to Free an Empire's Slaves* (Houghton, 2005)

Hopkins, Rob, *The Transition Handbook* (Green Books, 2008), *The Power of Just Doing Stuff: How Local Action Can Change the World* (UIT Cambridge, 2013)

Kenner, Robert, screenplay, *Merchants of Doubt*, film. Robert Kenner, Director (Participant Media, 2014)

Klein, Naomi, *This Changes Everything: Capitalism vs the Climate* (Simon & Schuster, 2014)

Langlois, Krista, "Following Ancient Footsteps," *High Country News* (10/2/2017)

Langston, Nancy, *Forest Dreams, Forest Nightmares: The Paradox of Old Growth in the Inland West* (University of Washington Press, 1995)

Laurent, Melanie and Cyril Dion, screenplay, *Tomorrow* [documentary]. Cyril Dion and Melanie Laurent, Directors. (Distributed by Mars Distribution, 2015)

Leahy, Terry, "The Chikukwa Permaculture Project (Zimbabwe)–The Full Story," *Permaculture Research Institute* (8/15/2013), permaculturenews.org

Lerch, Daniel, ed., *The Community Resilience Reader: Essential Resources for an Era of Upheaval* (Island Press, 2017)

Li, Qing, *Forest Bathing: How Trees Can Help You Find Health and Happiness* (Penguin, 2018)

MacLean, Nancy, *Democracy in Chains: The Deep History of the Radical Right's Stealth Plan for America* (Viking, 2017)

Martinez, Xiuhtezcatl, *We Rise: The Earth Guardians Guide to Building a Movement That Restores the Planet* (Rodale Books, 2017)

McDonough, William, and Michael Braungart, *Cradle to Cradle: Remaking the Way We Make Things* (Farrar, Straus and Giroux, 2002)

Merkley, Senator Jeff, "Create jobs without jeopardizing our future," *Mail Tribune*, 12/7/2017

Nadelson, Scott, "The Vulnerability of Outrage," *The Writer's Chronicle,* May/Summer, 2019

Robin, Vicki, *Your Money or Your Life: 9 Steps to Transforming Your Relationship with Money and Achieving Financial Independence* (Penguin Books, revised 2018)

Roosevelt, Eleanor, *Tomorrow Is Now: It Is Today That We Must Create the World of the Future* (Penguin, 2012, first published 1963)

Rush, Elizabeth, *Rising: Dispatches from the New American Shore* (Milkweed Editions, 2018)

Scott, Jared P., screenplay, *The Age of Consequences* (PF Pictures, 2016. Film)

Shuman, Michael, *Local Dollars, Local Sense: How to Shift Your Money from Wall Street to Main Street and Achieve Real Prosperity* (Chelsea Green Publishing, 2012)

Wallace, David Rains, *The Klamath Knot: Explorations of Myth and Evolution* (Sierra Club, 1983)

Wilson, Edward O., *Biophilia* (Harvard University Press, 2009)

Wood, Mary Christina, *Nature's Trust: Environmental Law for a New Ecological Age* (Cambridge University Press, 2014)

Index

About the Author

Having spent the last quarter century living off-grid in the foothills of Oregon's Coast Range with her husband David, Evelyn Searle Hess can study natural ecosystems right outside her door. A healthy ecosystem is diverse and complex enough to maintain its vigor, structure, and function through changing stresses. If it gets out of balance—overly simplified or differentially healthy—it loses resilience. So, too, do communities, whether of plants, animals, or people. As Hess watches human-caused deterioration of the biosphere—the global ecosystem—and increasing inequities and hardships in human communities, she is reminded that everything connects to everything else, an observation first attributed to Leonardo da Vinci.

Hess began her life in rural Washington State during Depression and World War years, giving a broad foundation in simple living. Her mother's fascination with the natural world and her attorney father's analytical mind helped her early to explore environmental, community, and life-cycle differences in plants, animals, and humans. Hess studied journalism, landscape architecture, and horticulture at the universities of Washington and Oregon and natural history throughout life, wherever she found herself. A writer, gardener, mother, grandmother, and great grandmother, she is the author of the award-winning *To the Woods: Sinking Roots, Living Lightly, and Finding True Home* and *Building a Better Nest: Living Lightly at Home and in the World*. Hess commits the remainder of her time on Earth to choices and actions that further social and environmental health nationally and beyond.

CPSIA information can be obtained
at www.ICGtesting.com
Printed in the USA
BVHW070113310321
603716BV00002B/2